PARRIS ISLAND

PARRIS ISLAND
ONCE A RECRUIT, ALWAYS A MARINE

EUGENE ALVAREZ, PHD

Published by The History Press
Charleston, SC 29403
www.historypress.net

Copyright © 2007 by Eugene Alvarez PhD
All rights reserved

Front cover: United States Marine recruits during close order drill. *Courtesy United States Marine Corps.*

First published 2007
Second printing 2010

Manufactured in the United States

ISBN 978.1.59629.292.5

Library of Congress Cataloging-in-Publication Data

Alvarez, Eugene.
Parris Island : once a recruit, always a marine / Eugene Alvarez.
p. cm.
Includes bibliographical references.
ISBN 978-1-59629-292-5 (alk. paper)
1. Parris Island (S.C. : Recruit depot) 2. United States. Marine Corps.
I. Title.
VE23.A95 2007
359.9'670975799--dc22
2007017129

Notice: The information in this book is true and complete to the best of our knowledge. It is offered without guarantee on the part of the author or The History Press. The author and The History Press disclaim all liability in connection with the use of this book.

All rights reserved. No part of this book may be reproduced or transmitted in any form whatsoever without prior written permission from the publisher except in the case of brief quotations embodied in critical articles and reviews.

This book is dedicated to all honorable Marine
Corps drill instructors and recruits.

―――――――――――

Esprit de Corps includes victory, teamwork, a feeling that Marines are a
special breed of people, and that each individual Marine, no matter what
his rank, is a self-appointed guardian of the revered reputation of
The Corps.
Leatherneck magazine, November 1942

CONTENTS

Preface	9
Prologue	12
Chapter One	15
Chapter Two	19
Chapter Three	24
Chapter Four	33
Chapter Five	45
Chapter Six	54
Chapter Seven	71
Chapter Eight	82
Chapter Nine	97
Chapter Ten	112
Chapter Eleven	125
Chapter Twelve	141
Chapter Thirteen	147
Contributors Living and Deceased and Credits	155
Suggested Bibliography	157
About the Author	159

PREFACE

The two Marine Corps recruit depots are among the most famous military bases in the United States. They have trained Marines for military service that is internationally known. To visit either the Parris Island, South Carolina or the San Diego, California installation is to relive much of American military history and to walk on revered grounds that have been known to Marines over the years.

Parris Island was a forbidding and unfriendly site of Navy gray barracks and metal Quonset huts when I first arrived at the recruit depot in August 1950. There was little opportunity for an eighteen-year-old high school dropout to learn much about the no-nonsense island's appearance, as the training was all work and no play.

The 1950–53 Korean War had begun and the two Marine recruit depots were hastily preparing for another Asian war. I survived boot camp, and my November 1950 orders were to report to the Second Marine Division, Camp Lejeune, North Carolina. Service followed in Korea with the First Marine Division in 1952–53. I became a Parris Island drill instructor in 1953–54. A second Parris Island tour was served from 1956 to 1959.

I learned much about Marine recruit training in the 1950s, and often considered recording my observations and those of others at some future time. But that ambition was delayed until being honorably discharged from the Marine Corps in 1959 to pursue a PhD and a college teaching career. My Parris Island interest again surfaced as a newly graduated professor. However, although a Parris Island recruit and a two-tour drill instructor, I did not yet possess the base historical expertise to record the process that transforms young men and women into United States Marines.

PREFACE

A wonderful opportunity came to me while attending a United States Air Force Historians meeting, at Warner Robins Air Force Base, Georgia. I inquired from my Air Force hosts about available funding to assist any Parris Island research, to include the evolution of recruit training and drill instructors, also known as DIs. My good fortune was realized with the receipt of a research grant from the Marine Corps Historical Center, Washington Navy Yard, Washington, D.C. Chief Historian Mr. Henry I. Shaw informed me, "Where could we have found a more qualified person to write a history of the depot and its recruit training than by a former Parris Island recruit and drill instructor with a history PhD."

By the time that I retired from teaching, my Parris Island research for the Historical Center was completed and I had produced several Parris Island books; but I had yet to publish any work that was exclusively dedicated to informing others about the basic training of former Marine recruits.

In August 2005, Mr. Larry Smith telephoned me for drill instructor information for use in his book, *The Few and the Proud*. Larry was a former editor for *Parade* magazine and I attended his Parris Island book signing in June 2006. I was impressed with the signing attendance and the interest in Mr. Smith's book that was mostly an oral history of Marine Corps drill instructors. My desire to produce a similar collection of past recruit and drill instructor lore emerged once more.

The approach to this work is historical, and "Old Corps" versus "New Corps" arguments are not intended. Opinions and recollections are freely given from qualified persons to include recruit and drill instructor experiences, and the offerings of historians, journalists and others who were and are informed with the Marine Corps' recruit training programs. Being the work addresses a chronological and historic recruit record, more emphasis is placed on earlier times rather than on more recent training.

Any inclusion of profanity is intended as historical record. Should anyone be offended, apologies are offered. But to omit profane records from the work cannot be acceptable.

This history could not have been completed without the assistance of others. I am indebted to the Marine Corps Historical Center for the Parris Island research grant that I received. I was not only awarded research expenses, but also given the unrestricted use of the Washington Naval Yard Center Library and the expert direction of the center's staff. Equally as important was the opportunity for me to meet Marines and others whom I could not have known before this undertaking.

The Parris Island command was always courteous to me while accommodating my base research. Majors Billy Canedo and Billy Moore kindly shared their time with me. Chief Warrant Office-2 Phyllis Alexander

PREFACE

(USMC Retired) was a dear and valuable former assistant until her untimely death. I am eternally grateful to Phyllis. Judge John C. Stevens III proved to be a loyal friend and an excellent critic and supporter for me, as he has been in the past. Parris Island Museum Director, Stephen R. Wise, PhD, assisted me regarding significant photographs and recent Parris Island events. Without his attention I would have missed numerous projects and personal opportunities that came to me via Dr. Wise. Thank you, Steve.

I am also indebted to the World War II Parris Island drill instructors who so enriched this book. First Sergeant and former drill instructor Charles Carmin (USMC Retired) critically read the text. Former Marine Mr. Don Mason offered text suggestions and poetry for the book. My gratitude is extended to Mr. Larry Smith for listing this author as a drill instructor in *The Few and the Proud*. Larry's published study inspired me to at last undertake this work.

Lieutenant Colonel Tom McKenney (USMC Retired) contributed unknown text and photographs, while Major Joe Featherston (USMCR Retired) gave a detailed account of his 1956 recruit experience. Major Featherston was kind enough not to implicate his former senior drill instructor of violating any training rules. Thank you, Major, sir! Ms. Joan Redenbaugh was always supportive of this work as a stylist, photograph organizer and a proofreader.

My gratitude is also offered to my editors, Ms. Deborah Carver and Ms. Jenny Kaemmerlen.

This author is further indebted to other contributors, who are too numerous individually to list. A broad listing of credits follows the story text and apologies are extended to anyone who was inadvertently overlooked. Copious text notes and a lengthy bibliography have been omitted from this work for reading ease and comfort. A select and brief listing of books is offered.

It should be noted that the author never visited the San Diego depot nor researched that base history. Hence, this story leans heavily toward Parris Island. Nonetheless, former recruits and San Diego drill instructors have enlightened me regarding the equally important California recruit training base.

PROLOGUE

Marine Corps recruit training has evolved over the years, and no two recruits are alike. Each Marine has a story to tell. Yet the training mission remains the same: to instill esprit de corps, discipline and obedience to orders, and to impart into young civilians how to function as a fighting team. Marksmanship and physical training are essential to the recruit's instructions, too.

The reliance on physical discipline that was once a part of Marine boot camps is no longer allowed. Whereas mental duress was previously a significant part of the curriculum, less stress is desired for recruits in recent times. Succinctly, harsh training methods that were accepted in the "Old Corps" can or may be grounds for a court-martial today.

Marines train for war and the Corps is eager to be the "First To Fight." But that pledge cannot be an idle boast. Much that appears in this book as harassment and cruelty was, in times past, viewed as preparing for war. When the ramp of a World War II landing craft fell on the Pacific beach of an enemy-held island, there was no time for a debate. Discipline, obedience and marksmanship were paramount.

Drill instructors schooling was largely an "on the job" undertaking until a Marine Corps Drill Instructor School was created for each of the two recruit depots in 1952. Until more recent times drill instructors enjoyed near complete authority and resorted to a sundry of personal techniques and training strategy with their ever loyal and attentive recruits. Parris Island's "Locker Box Jones" seems to have utilized the heavy and green cumbersome wood recruit footlockers for training and in barracks close order drill. Other drill instructors simply used a powerful voice that could turn the air blue. One recruit ruefully dreamed:

PROLOGUE

If an angel came to me,
And, asked what I would like to see,
There could only be one choice—
A DI who has lost his voice.

Young men and women were and are attracted to the Marines for sundry reasons. Many have the desire to serve with "the best." A romance gone sour sent both men and women to the Corps' boot camps. The dress blue and the smart green uniforms attracted others to abandon an unhappy home life or a small town environment that offered little or no challenging life.

Many desired to wear the Eagle, Globe & Anchor, or were inspired by a motion picture. The personal challenge and acquisition of confidence that accompanies severe and demanding training or the desire to follow in the footsteps of a former relative Marine motivated others. Patriotism, personal advancement, the challenge to make the grade and that Marines are taught to never let another Marine down inspired many more to enlist.

A crusty and veteran Marine major once offered that, "With liquor and youth, one could conquer the world." Perhaps the major was correct. But rigorous training and Marine Corps discipline surely helps. More realistic are the comments of Major Gene Duncan (USMC Retired) who advanced through the ranks and was wounded in Vietnam. The prolific author offered:

> *The purpose of boot camp should be to instill the discipline which will cause that Marine to stand and fight when every inborn instinct tells him to run.*
> *Mental endurance—the ability to effectively function under great pressure—is every bit as important, if not more so, than physical endurance.*
> *Life should—must—be made miserable for the recruit.*
> *I would not be willing to take into combat any Marine who ever accused his DI of anything short of murder.*

The Marine esprit de corps seldom, if ever, dies. Harold Kenneth Seymour enlisted in the Marines at seventeen years of age, standing five feet eight inches and weighing 118 pounds. Drill instructors called such lightweights "Feather Merchants." Seymour's Post Honor Platoon 351 graduated from Parris Island on December 1, 1953, under the command of drill instructors Corporal R.R.Grant and PFC T.F. Allen. Drill instructors are not easy to forget.

Private Seymour advanced to the rank of a Marine sergeant with six years service. An adventurer, Seymour joined the Air Force for three

PROLOGUE

years, and then entered the United States Army. The former Marine was commissioned an Army officer and served with the famous Big Red One Division in 1966 in Vietnam. A second 1971 Vietnam tour saw Seymour commanding four hundred men assigned to Task Force Alpha at Vinh Hao and Don Duong.

> Colonel Seymour recalled, "My Marine Corps Boot Camp training at Parris Island gave me a solid foundation for this unusual command. I remember telling a very large formation of my men, who were griping about serving in Vietnam, that I'd willingly stay in Vietnam for ten years before I'd go back to Parris Island for ten weeks."

The officer spent twenty-eight years in the United States Army and was "airborne all the way." Colonel Seymour attends a yearly Marine Corps Birthday Ball wearing his Army mess dress uniform, but never forgetting Parris Island and the Marine Corps. "Semper Fi! Once a Marine, Always a Marine!"

Chapter One

"WHERE IT ALL BEGINS"
PARRIS ISLAND AND SAN DIEGO

Marines have historically compared the two recruit depots and argued which base was the most demanding. Parris Island's boots offer the South Carolina heat and insects and the former sand parade field that was used for maneuvers and to train recruits. San Diego recruits are branded "Hollywood Marines." Conversely, the California boots list the San Diego noise and training distractions and their hikes over the Southern California hills. Mr. L.F. Ruttle had the dubious, yet rare opportunity to experience both depots at contrasting times in the Roaring Twenties and during World War II:

Parris Island (1928)

Liberty: None (too far away)
Chow: Nourishing
Weather: Lousy (sand fleas added)
Roughness: Mayhem-by-design
Barracks: Not the Waldorf-Astoria Hotel
Training: Excellent
DIs: Lovable monsters
Location: Far from home

San Diego (1943)

Liberty: None (and no place to go if you had it)
Chow: Nourishing

Parris Island

Weather: Lousy (cold, fog added)
Roughness: Mayhem-by-design
Barracks: Not planned by Hilton
Training: Better
DIs: Educated lovable monsters
Location: Far from home

Mr. Ruttle added, "I'd send the South Carolina sand fleas to San Diego for a short hitch and the California fog to Parris Island for a time. Alternate them every year and end the argument about which one is rougher once and for all." The fact is that both of the bases train those who are among the finest combatants in the world.

PARRIS ISLAND

Although Parris Island, South Carolina, is famous for its heat, sand gnats and rigorous training, it is often overlooked that the island's history is significantly important. The base and its coastal vicinity were inhabited by the Indians, French and the Spanish until the English arrived. The coastal towns of Charleston and Beaufort later emerged. In 1715 Colonel Alexander Parris purchased land on the South Carolina island that now carries his name. Island slaves worked on plantations until the American Civil War.

As the United States advanced toward world power status, Parris Island and its adjacent waterways in the late nineteenth century attracted a major southeastern naval base. The U.S. Naval Station Port Royal (Parris Island) was dedicated on June 26, 1891. A massive dry dock was completed in 1895 that welcomed the largest naval war ships to the wooden facility. The impressive dock survived a monstrous 1893 hurricane that killed several thousand persons along the lower South Carolina coast. However, the Parris Island dock was defenseless to political storms. The facility lost its importance to a Charleston, South Carolina installation and was closed in 1909. The mud-filled dock remains evident today.

The naval station played a minor role in the 1898 Spanish-American War. Parris Island briefly welcomed an officers' training school and a short-lived recruit program that was lost to Norfolk, Virginia. Parris Island was additionally designated a military prison. The Navy relinquished the Port Royal Naval Station (Parris Island) to the Marine Corps. Marine Barracks Port Royal was designated on October 25, 1915, as Parris Island recruit training began.

CHAPTER ONE

SAN DIEGO

San Diego's early history is associated with the Spanish and Mexican possession of the Southwestern United States. Lands that included California were gained by the United States during the Mexican War, 1846–48.

West Coast Marines first trained at Puget Sound, Washington, and the Mare Island, California Navy Yards, until attention was focused toward a San Diego base around 1914. Construction was in progress by 1919, and much of the naval base was built on a filled tidal marsh that was adjacent to San Diego and an infant city airport. It appears that the earliest base buildings were dedicated in 1921. The facility began training recruits by 1923.

Being that the San Diego Marine Corps and Navy bases are adjacent to one another, harrowing experiences have been known to arriving Navy and Marine recruits. When twenty new 1957 recruits arrived at the San Diego Marine Corps Recruit Depot, one young man frantically attempted to attract the attention of the Marine drill instructor. The recruit was reprimanded, but the sergeant finally listened to him. "Sir, I'm supposed to be in the Navy!" It was discovered that eighteen of the twenty recruits were U.S. Navy enlistees who had boarded the wrong bus. They were taken to the Naval Training Center nearby.

The San Diego post was strategically important to the World War II Pacific theater, and a number of motion pictures used the California depot and other military bases nearby for on-location sets. Three of the five Marines (and one U.S. Navy Hospital Corpsman) who are pictured in the famous 1945 Iwo Jima flag-raising photograph were San Diego recruits. San Diego receives about half of the Marine Corps recruits other than female Marines, who are trained at Parris Island.

CURRENT RECRUIT TRAINING

Recruit training is mostly uniform at the Parris Island and San Diego depots. It is noted that the following narration is a generalization and does not go into much of the detail that is currently significant to boot camp.

New recruits are presently bussed to the bases or flown to nearby airports, and then transported to the boot camps at night. The one week of forming and twelve weeks of training are generally divided into three phases. Still, there are slight variations between the California and South Carolina recruit depots. It is again noted that all female recruits train only at Parris Island.

The recruit receives an M16 rifle, an initial clothing issue and the opportunity to telephone a family or friend to inform them that they have safely arrived at their destination. The recruits are assigned a barracks squad bay and bunks and are soon introduced to their drill instructors. Strength tests are conducted, and there are administrative and medical details. Recruits are permitted to attend church call.

Martial arts instruction commences during the early training days, as does pugil stick training. Both physical programs are intended to build confidence, and the safety of the recruit is always a major concern. Other early training includes Marine Corps history classes and close order drill.

The sixth to the eleventh training days add the running of the confidence course, which includes the Slide for Life, the Sky Scraper, the Belly Buster, the Tough One and the Dirty Name. The gas chamber is visited during the twelfth to the seventeenth training days. The annoying agent is similar to tear gas. Through the twenty-third training day more physical and classroom schooling continues, as the recruits begin to master drill and to work as a unit team.

Phase two, from day twenty to day forty-one, introduces swimming qualifications. Every recruit must pass the demanding swimming tests to graduate from boot camp. Drill and class instructions continue that includes a first written exam. Other previous mentioned training continues in the active schedule.

Phase three covers training days forty-two to sixty. More medical and dental evaluations are administered. Field exercises include the Basic Warrior Training for recruits and marksmanship training at the rifle range. Much San Diego field activity and all rifle training are conducted at Marine Corps Base, Camp Pendleton, which is some forty-five miles north of San Diego. The famous base has also been the home of the First Marine Division.

A final fitness test is administered prior to challenging the Crucible. That event is physically demanding and is designed to promote teamwork and leadership. Upon completing the Crucible, recruits share a "Warrior's Breakfast," which is "an all you can eat buffet shared with the drill instructors."

Week twelve is devoted to more administrative details. Inspections are held and the recruits' training issue and rifle are returned. Pinning on the Eagle, Globe and Anchor emblem identifies the recruit as a Marine. A graduation parade marks the end of recruit training and is attended by hundreds of families and friends. Advanced field training ensues following a boot camp leave, at the Marine Combat Training Schools.

Chapter Two

IN THE BEGINNING WAS THE LEFT FOOT.

T. GRADY GALLANT

It would seem that the first American Continental Marines trained at their Philadelphia birthplace in 1775. The training was most likely austere, and obedience to orders was paramount. For example, it was recorded that a Marine private found sleeping at his post in the very, very "Old Corps" was sentenced to walk his post for two months, while "encumbered by a ball and chain." Another Marine was court-martialed in 1820 for desertion and sentenced to "wear an iron collar, with a six-pound shot attached, for four months." He was also ordered to forfeit all pay and to be "drummed ignominiously" out of the service. Others were lashed with a cat-o'-nine-tails whip or deprived of their rum allowance. Some Marines were made to "drink one or two quarts of salt water" as punishment for drunkenness.

Being that Marines were and remain a significant part of the United States Navy, recruits were schooled at naval yards before and after the 1861–65 American Civil War. John C. Ogle was attracted to the Marines' colorful uniforms that influenced him to enlist in 1878. Ogle's brief recruit training at the Washington, D.C. Navy Yard consisted of learning to drill and to follow orders.

Marine basic instruction at the Brooklyn Navy Yard lasted about ten days during the 1890s. The succinct program included close order drill, and the mastery of other skills that were grasped by the recruits as best they could. A private's pay was $12.80 a month. A brief three-company recruit program existed on Parris Island in 1911. Few recruits trained on the South Carolina base until later years. West Coast recruits as noted trained at the Puget Sound, Washington, and the Mare Island, California Navy Yards, until the post–World War I completion of a San Diego base.

Parris Island

Charles B. Hughes arrived on Parris Island in 1911 to be a prison guard and later became a drill instructor. He discovered that some recruits did not learn as quickly as others, and that they were placed in a "coordination platoon." Hughes requested to train these platoons, since he was convinced there "was no such thing as an inept recruit," and that a good instructor could turn slow learners into Marines. Hughes argued that it depended on the instructor, who could "make or break a platoon on its first training day."

Twenty-year-old O.T. Cox was impressed with twelve-foot Marine recruiting posters that influenced him to enlist. The young man shipped out for Mare Island, California, on December 17, 1905. After four or five days, Cox was placed in a recruit squad and issued uniforms. They included two suits of blue overalls. One set was reserved for inspections and the other was a working uniform. Two suits of heavy woolen underwear, four pairs of socks, two pairs of shoes, two flannel shirts and an overcoat were issued in addition to two blankets. His equipment consisted of a canteen, a canvas haversack, a Krag-Jørgensen rifle and a bayonet.

For three days Private Cox learned "foot movements and turns." Two additional days were devoted to the rifle manual of arms, as Cox was graduated from his basic training status to a regular company. Other recruits were schooled to serve on U.S. Navy ships or to be naval military prison guards.

Norfolk was the largest Marine recruit training base until the Parris Island depot replaced the Virginia facility. Newly arrived 1914 Norfolk "applicants" were immediately required to bathe and were assigned a bunk. Their civilian clothing was fumigated and sterilized, and military pajamas could be a recruit's first uniform. An applicant had the option of whether or not to enlist before any training commenced. But during the interim one had to bathe daily, make one's bunk and conduct an early morning clean up or "police call." According to recruiting literature, applicants received free smoking and chewing tobacco, access to magazines and card games and the use of a music machine.

The Norfolk applicants were administered examinations during their second and third days to determine if they were fit for military service. The men were kept under surveillance at all times and treated in such a fashion as to dispel any romantic notions about military life. Rejected applicants were mostly refused enlistment because of misconduct, for psychological reasons or "racial instinct and home sickness." The system apparently worked, since only a small percentage of the Norfolk men were "elopers" (deserters) or refused to enlist.

Once the USS *Prairie* discharged its cargo of Norfolk Marines and recruits at Parris Island on October 21, 1915, construction for needed

CHAPTER TWO

facilities immediately began. Additional quarters and a larger rifle range were required and a main boulevard was cleared. The established naval station was mostly near the waterfront. An adjunct to the isolated base was the Yemassee railroad junction some twenty or more miles away. Recruits were at this time transported from the mainland to Parris Island by boat.

A first assignment for new arrivals was to have a full meal at the receiving barracks and to receive an issue of soap, towels and pajamas. After showering in hot or cold, fresh or salt water, the recruit was assigned a bunk that must be made according to regulations at all times. As at Norfolk an applicant still had the option of declining enlistment during his first days on Parris Island.

Reveille was sounded at 0600. Applicants washed and shaved before the morning meal, policed their areas and made their bunks. A series of physical and mental exams followed including a "dynamometer" test. This machine automatically registered "the exact number of pounds pulled by each muscle group of the body." The machine was used during training to record the physical development of recruits, whose growth was "surprising in most cases." For instance, in 1916, the Ninth Drill Company of fifty-six men averaged a weight increase of seven pounds per recruit during fourteen weeks of instruction.

After a recruit's tests were passed an applicant was sworn in and met his instructors who trained the boots for a normal fourteen-week course. The program included

Care of Clothing and Person	Military Courtesies
Carrying Messages	Physical Drill
The Rifle Nomenclature	Manual of Arms
Squad and Company Drill	Individual Cooking
Packing Knapsacks and Blanket Rolls	Bayonet Exercises
Pitching and Striking Tents	Guard Duty and Street Riot Drill
Field Formations and Fortifications	Wall Scaling
Boxing	Athletics and Swimming
First Aid	Handling of Boats

Water discipline was most likely stressed since Parris Island's sparse supply had to be boiled and was unpleasant to taste.

By the sixth or seventh training week, the recruits marched to their rifle range quarters, where they lived for several weeks. Recruits "snapped in," or dry-fired their rifles, until live ammunition was issued to them. The training was to qualify each man with the Model 1903 Springfield rifle as a marksman, sharpshooter or an expert. An expert's medal was awarded by posting 252 out of 300 points. A sharpshooter was required to earn 237 points, and 201 points was required to become a marksman. Two, three or five extra dollars was added to a Marine's monthly pay, depending on one's rifle score.

Recruits returned from the rifle range to the main station upon the completion of their marksmanship training, to resume the course of instruction previously described. The Parris Island mental rigors seem to have been reduced at this time, since the *Recruiters' Bulletin* stated that available recreation for recruits included bowling, pool and athletic competition with cash rewards. Nightly motion pictures were shown and a library was available. The *Bulletin* reported that "fruit, milk, tobacco, candy" and other items could be purchased at the Parris Island Post Exchange (PX). The magazine added, "Everything possible is being done for the entertainment and amusement of the men at the post." Recruit recollections do not support such glowing appraisals of boot camp comforts.

Marine recruits were designated "boots" in Norfolk and most likely elsewhere. The origin of the word is uncertain. One argument is that "boots" was derived from the canvas leggings that Marines wore as late as circa 1952. A more plausible argument is that during the early 1900s, all naval recruits were issued rubber boots to be worn while cleaning the decks of ships. Regular Marines and sailors wore out or discarded their rubber boots, making the newly arriving men more conspicuous. Hence, arriving recruits with new rubber boots were welcomed on board a ship with the greeting, "Here come some more rubber boots." Through the years the appellation was simply shortened to "boots."

Parris Island's earliest "boots" not only experienced rigorous training, but were also indoctrinated with a 1916 depot publication titled *Semper Fidelis*. Its purpose was to "instill into the recruit that esprit de corps, that feeling of loyalty to the Marine Corps and to the flag that has been the secret of our glorious record, and which is so essential to efficient service." Other lessons and indoctrination were provided by lecturers who addressed recruits about "why I am proud to be a Marine," Marine Corps history lessons and the recognition of Medal of Honor recipients.

Private Homer A. Throop trained at Parris Island in 1915 with the first and initial command of base recruits. The boot enlisted in September and first reported to Norfolk. After Throop was in Virginia about three weeks the word was passed to "pack your bags, we're going to Parris Island." Throop recalled that he spent about three days at sea on the USS *Prairie*, and until the ship anchored off Parris Island in the bay. The men were ferried to a Parris Island beach in a motorboat and were assigned to live in tents that were pitched in the vicinity of the wooden dry dock.

CHAPTER TWO

At seventy-seven years of age, the elderly but spunky Marine returned to Parris Island in 1969, to witness his grandson graduate from boot camp. The only physical and natural objects that Throop recognized were the defunct dry dock, several buildings and the South Carolina marsh. Throop stated, "I'll bet the mosquitoes are just as big and the sand fleas are still as pesky as they have always been."

Three others that trained at Parris Island in 1916 were Privates Elvis Campbell, Samuel Winfield and Oliver Trombly. Campbell was scheduled for twelve weeks of schooling and discipline, but his instruction was reduced by orders to join a Santo Domingo expedition. While Campbell was on Parris Island he helped to unload coal barges and "the first barge of oyster shells," which were used for base road construction and landfill.

Private Winfield arrived in South Carolina in 1916, and was "immediately hustled aboard a raft" that was "hand-pulled to Parris Island." Following his examinations and boot camp issues, Winfield began his military career at $14.80 a month. The recruit quarters included tents that were enclosed by barbed wire.

The Parris Island day began at 0500, with a later breakfast of coffee, bread and beans. The day was secured at 2000 (10:00 p.m.) Winfield and others vividly recalled the rawness of the isolated depot and saw only one base automobile. Private Winfield remembered that recruits followed a trail to the rifle range that was inhabited by snakes and an occasional alligator.

Private Trombly enlisted for four years and remembered there were about six hundred Marines on Parris Island, who were commanded by a major in 1916. The lad was assigned to a tent and was issued a rifle, a cartridge belt and a pack. His uniforms included one green wool uniform with a high choke collar, spiral puttees and two high-collared khaki uniforms. Trombly's training was mostly drill. However, during his six weeks on Parris Island, Trombly's platoon went for a daily morning swim in the Beaufort River.

Even though Trombly considered his training "a little tougher" than that of future recruits, he was grateful that it was demanding before he faced a German Army as a Marine "Devil Dog" during World War I.

Chapter Three

PARRIS ISLAND IS NOT THE KIND OF PLACE
YOU FORGET.

MAJOR PERRY N. COLEY

Parris Island first accommodated a major recruit overload during World War I. The tremendous growth resulted from the passage of National Defense Acts, and the Marine Commandant recommending a Corps growth from 7,200 to 10,000 men. The South Carolina depot eventually trained more than 46,000 World War I Marines.

Some Marines debate that the First World War marked the end of the "Old Corps," as California training centers and Parris Island prepared Marines to fight in France. It is also argued that the World War I recruits endured more strenuous training than any others have on the Parris Island base.

The process began at the Yemassee rail junction and continued via the railroad to the town of Port Royal. The new arrivals were conveyed to Parris Island by boat or barge and then quarantined. Recruits frequently washed their clothing in salt water on wooden racks, or even on a Parris Island beach. *Living Age* magazine reported that cleanliness and personal hygiene were stressed to perfection. The Marines' "two salient characteristics" were their "ability to make something out of nothing and to do it quickly" and the distinction that when they were not the first to fight, they were the "first to clean."

Following clothing issues and examinations the new boot was transferred "over the fence" from quarantine to train from four to fourteen weeks. The length of time depended on the urgency for more men in France. Under normal conditions the program was divided into three phases. The initial instruction mostly addressed drill and schooling in basic military subjects. They included the general orders, military courtesy, pay call procedures, first aid, weaponry and always drill.

CHAPTER THREE

Marksmanship was the second training phase. One officer wrote that the World War I recruit "learned the Marine way from some hard-bitten professionals who valued discipline and a steady trigger finger above all else." The proper care of one's rifle was always stressed:

> *Your rifle is your best friend; take every care of it.*
> *Treat it as you would your wife;*
> *Rub it thoroughly with an oily rag every day.*

The third phase was spent in the "maneuver" areas. Bayonet drill was practiced, field problems were conducted and entrenchments were dug, making the training area resemble Europe's Western Front. Construction details were assigned and one World War I recruit recorded, "We built Parris Island while we were going through boot camp."

The urgent base expansion, along with strenuous training and Parris Island's isolation, marked the beginning of a physical and mental type of Marine Corps instruction, which later acquired international respect. It was argued that the reason why the Marines "performed with such uniform excellence in France," was attributable to the World War I rigors of Parris Island, "where, despite the terrible pressure of war, not an ounce of essential instruction was stinted and not a scintilla of the traditional individual discipline [was] omitted."

Future Lieutenant General Merwin H. Silverthorn observed Parris Island's severe training after reporting there following his Mare Island boot experience and Quantico, Virginia officer training: "We could hardly believe the stories about Parris Island, and especially those of recruits scooping, bare-handed, buckets of oyster shells to build roads and for land fill use."

Another significant and unique training aspect was the augmented authority of the drill instructor. Some attribute that to a strong Prussian influence that infiltrated the Marine Corps. It is more likely that between 1915 and 1917, the depot policy was to organize boots into standard rifle companies that were commanded by officers who were assisted by noncommissioned men. But "somewhere between the planning and the execution stage," the Marine Corps discovered that its limited officer numbers were insufficient for the rapidly growing number of recruit companies and the prescribed commissioned officer berths. Much of the authority reserved for commissioned officers was passed down to the enlisted instructor, with officers more, but not exclusively confined to administrative details.

It was not long until the enlisted sergeants established "commensurate authority" over their recruits. Recruits were taught by their drill instructors that the Marines did everything according to rank. At Parris Island, "the DI was junior only to God." Newly graduated recruits were appointed as instructors, or "Acting Jacks," in times of need. Many outstanding boots numbered future generals who began their careers as enlisted men. The officers never forgot their recruit experience or drill instructors, and mostly supported any future and similar training that they had endured.

One of many future generals who trained at Parris Island was Lieutenant General Lewis B. "Chesty" Puller, who enlisted in 1918. Following six days in quarantine, Private Puller was placed under the charge of a "towering Paris-born Dane." Corporal John DeSparre spoke six languages, was a student of history and apparently had an eye for leadership. Puller was appointed to head a platoon in his recruit company as early as his third training day. Later, upon graduation, Puller was ordered to the base noncommissioned officers' school to receive some drill instructor training under a "spectacularly profane and ungrammatical" old Marine captain. The officer taught Puller bayonet, rifle, drill, judo and boxing skills, which General "Chesty" Puller used with consummate skills during his legendary Marine Corps career.

Major General Melvin L. Krulewitch vividly recalled his 1917 DI was a semi-illiterate Czechoslovakian who appointed him to be an "Acting Jack." That entitled Krulewitch to wear a leather belt and to mingle with noncommissioned officers (NCOs). Parris Island was recalled that "except for a brick building in the main headquarters, the place was nothing but shacks and tents." Recruits spent five to eight days in an "applicant's camp," until venereal or other diseases surfaced. Then "we went across the line" and "into the boot camp." That environment was a "totally different world," where others hurled obscene phrases at the new recruits: "It was so filthy and dirty."

Krulewitch's first meal was "greasy fatback, side meat, with blackstrap molasses, and black coffee. Our bedding was crawling with bugs." The recruits came from all over the country. Many chewed tobacco and rolled their own cigarettes. There were numerous fights. On every other training day the boots hauled oyster shells for building base roads. Bricks unloaded from barges left bleeding fingers and hands. Steel beams weighing several tons were also unloaded from barges at the base dock. Krulewitch recalled that he never knew training as severe as at Parris Island, and people said the "Foreign Legion and other military units couldn't compare to the early Marine Corps training that we had there."

Generals William W. Rogers and Gerald C. Thomas experienced a similar potpourri of instructors and recruits. Parris Island was remembered as "pretty grim" to Rogers, who was fed "wormy grits." Training was

CHAPTER THREE

done in the maneuver area and there was always drill. The recruits carried oyster shells in large buckets to construct base roads. Rogers impressed his instructors and was appointed an "Acting Jack." That recognition permitted the future general to eat at the sergeant's table.

General Thomas also enlisted in 1917. He was trained by a veteran gunnery sergeant named Borden and another semi-illiterate NCO who memorized the *Landing Party Manual*, from which he taught drill. Much of the drilling was done on sandy flats near a beach. The sergeants were "tough," but Thomas observed that the recruits were not physically abused since the DIs had "other ways to make us miserable," and "they turned out some good Marines."

The "other ways" took many forms, from scathing orders such as "Snap out of your hop, young fellah," to barrages of profanity directed at the boots. Receiving a perfumed letter could result in a recruit being ordered to find a "thousand feet of skirmish line," or to fetch a nonexistent key to the maneuver grounds. Physical repercussions might include a crack on the head with a swagger stick. If a DI "felt a little bad" on a particular day, a platoon might be marched from one end of the base to the rifle range while wearing gas masks. On other instances recruits were marched barefooted across oyster shells.

Other Parris Island trained general officers were Major General Ion M. Bethel, General Christian F. Schilt, General Ray A. Robinson, and Major General Walter G. Farrell. Bethel arrived on Parris Island via a barge and was greeted with chants of, "You'll be sorry. You'll be sorry," a greeting used widely during World War II. But as World War I ended, Bethel was assigned to work in a machine shop until he was discovered loafing on the job. This violation led to his being transferred to labor in a locker box factory until Bethel departed the depot for more challenging future goals.

When General Schilt arrived at Parris Island in June 1917, his civilian clothes were confiscated, and Schilt was showered and uniformed. Many recruits slept in tents, and as the future pioneer of Marine Corps aviation and Nicaragua Medal of Honor winner recalled, the training day began at 0430. The drill instructors were strict. Training included physical exercises, drill, studies and the seemingly ubiquitous labor of hauling oyster shells in buckets, which strengthened a recruit's arms and legs.

General Robinson reported to Parris Island and was issued a rifle, a hat (cover), shoes and two pairs of white pajamas, in which Robinson trained until his regulation shirts and trousers arrived. Since there were no maneuver area flush toilets, recruits had to utilize outhouses that were built over a creek. These were furnished with two logs on which one sat; these were worse than the crude facilities that Robinson saw in China in later years. General Robinson also recalled that whenever recruits received packages from home, the contents were guarded from thieves by filling the box with Parris Island land crabs. Boots carried more than twenty-three buckets of oyster shells daily over the loose sand. Although he spoke well of his drill instructors, Robinson summed up his training exclaiming, "Oh, my God, yes, it was really terrible."

Major General Walter G. Farrell recalled in 1917, the essence of the boot camp experience:

> *We had a sergeant DI and two corporals who took care of us day and night. They were tough but fair. They carried swagger sticks made of swab handles, and they used them frequently. We felt them across our legs, in the stomach or, worst of all they would drag the tip across your ribs like rattling a stick on a picket fence. But they always had my welfare at heart and by the time they were finished with me, I knew the meaning of instant obedience.*

Other officers who were not recruits recalled Parris Island. Second Lieutenant and future Commandant Lemuel C. Shepard Jr. found Parris Island a "deadly place" in 1917. Major General Omar T. Pfeiffer was unimpressed with what he observed during the formal parades and the manner in which persons were being commissioned at the base. Future Lieutenant General James L. Underhill was ordered to Parris Island to shepherd recruits destined for New York and Europe to find "the most horrible assembly of troops to go to war that one could imagine." To make matters worse, the officer was quarantined in a "concentration camp out on the sand dunes," which was surrounded by barbed wire.

More recollections addressed quarantines, cold saltwater and freshwater showers, the heat, wood-burning stoves and sandstorms that obscured targets on the rifle range. Strict discipline was always in effect. For talking back to an NCO, a recruit could receive a commonly imposed five days' rations of water and bread, popularly called "piss and punk." Yet as demanding as the training was, recruits seldom harbored grudges against their instructors and valued their difficult and arduous recruit training once they entered the frontline trenches in France.

Private Robert Leroy Strong was such a man. Strong was sworn into the Corps at Parris Island in December 1917, and spent his first week in the quarantine station. Strong was instructed in drill, rolling and marching with the heavy pack, listening to and viewing illustrated venereal disease lectures and conducting barracks-cleaning "field days," when one hundred buckets of water were used. His training routine began at 0500 and lasted

CHAPTER THREE

as late as 2300 and the recruits were kept occupied every minute of the day. Like others, Strong was homesick and especially missed his fiancée. His only solace from his "torture" was to attend base church services that were available to recruits.

Private Strong's platoon marched two miles from the main station to the rifle range and returned again for a noon meal. The march was repeated in the afternoon, making a total of at least eight miles of marching per day. The men washed their clothes in the evening, shaved, cleaned their weapons and did other chores. Rifle firing was at the two-, three-, five- and six-hundred-yard lines and the maximum score was 300 points. Experts were required to post 253 points; sharpshooters, 238; and marksmen, 202. Marksmen first- and second-class had to fire scores of 177 and 152 points.

Strong was transferred to "the overseas camp" at Quantico, Virginia, following his boot training, to embark for Europe. In France he fought at Chateau-Thierry, Soissons, Belleau Wood, St. Mihiel and in other brutal engagements as a machine gunner in the Eighty-first Company, Sixth Machine Gun Battalion. Strong always demonstrated his belief in the superiority of the Marines, maintaining that a Marine with two months of Parris Island training was equal to any Army soldier with two years' service. His free "Soldier's Mail" envelopes were always altered to read, "Marines' Mail."

Bob Strong was a Marine for only nineteen months. For the remainder of his life he was proud of his war record and for having been a World War I Marine. His gravestone reflected this pride, reading:

WORLD WAR
Robert L. Strong
U.S. Marine Corps
Died July 1, 1955
Age 60.

Private Carl A. Brannen enlisted at Parris Island in February 1918, and was assigned to a recruit platoon of about forty men for ten training weeks. His drill instructor was dissatisfied with anything short of perfection, and on one instance an instructor became so enraged the DI broke his swagger stick. The profanity turned the air blue. The junior drill instructor was as demanding. Private Brannen was on another occasion ordered to wash his socks a second time. Thinking they were clean and the order was unfair, the recruit was angered. Realizing that a boot had little appeal and was expected to obey, Brannen resigned himself to the task during training that "was much worse than expected." Prior to leaving the depot "through blurred eyes," each recruit contributed fifty cents to the drill instructors, who were last seen meeting another forming platoon of newly arrived recruits.

William Manchester Sr. spent less than four weeks on Parris Island, mostly building roads. At some point he qualified with the rifle and soon departed for France. Manchester joined the Fifth Marines and fought the Germans at Soissons and St. Mihiel. The Marine was horribly wounded in the shoulder and was without the use of an arm until his death in 1941. Manchester's son was a World War II Marine and a celebrated author.

Louis Lange drilled for eight or nine hours daily in "what seemed like 130 degrees." The standard service drill manual was used, and Lange spoke highly of his drill instructors, who he declared were equivalent to a full general in boot camp.

The bunkhouses, palmetto trees, dirt roads and the sand parade field left other memories. During physical examinations the men were identified by iodine painted numbers on their chest. Others recalled rope climbing, boxing, bayonet training, etc., and the spirit of the football field on the maneuver grounds. Recruits who did not know their left foot from their right were placed in the "booby squad" to be instructed in drill.

World War I recruits recalled bugle calls, orders shouted through megaphones, and food as tough as "turkey buzzard." Marine Corps haircuts were memorable, as were the Parris Island bedbugs that caused the recruits to wear their socks to bed at night to protect their feet. Parris Island malaria was also known. Underlying all of the above was the strict discipline that has always been characteristic of Marine boot camp. The penalty for failing to follow commands included being ordered to haul buckets of oyster shells, to double-time carrying one's rifle at port arms or overhead, or to be physically corrected.

The World War I recruits were elated "to get off the hell-hole called Parris Island." But each "Devil Dog" had the satisfaction that he had endured a physical and mental training program for which he could be proud. He knew that he and his fellow Marines could defeat their World War I enemies as United States Marines.

CHAPTER THREE

First and Second Parris Island Recruit Training Battalions, 1947.

Parris Island Parade Field, 1950s.

Wake Boulevard and Weapons Training Battalion, 1950s.

The Boulevard de France and parade field, 1975.

Chapter Four

WHEN I FIRST GOT HERE [PARRIS ISLAND] I THOUGHT I WOULD DIE. AFTER TWO WEEKS I HOPED I WOULD DIE. LATER I KNEW I WOULD NOT DIE BECAUSE I HAD BECOME TOO TOUGH TO KILL.

ANONYMOUS WORLD WAR I RECRUIT

The 1920s link from civilian life to the Marine Corps remained at Yemassee. The railroad junction saw the construction of a first-class drugstore and new brick buildings. Yet once a base causeway was completed, most Yemassee recruits arrived on Parris Island via barges or on a powered "kicker" motorboat. Parris Island's wartime and ensuing growth populated the depot more than the nearby towns of Beaufort and Port Royal combined. The largest number of 3,500 recruits arrived on Parris Island in 1920. The smallest number of 1,500 boots arrived in 1922.

The depot's peacetime atmosphere caused some to offer that the Old Corps was history and that recruits in the 1920s were a cut below those of previous years. One old-timer commented the days of receiving a rifle and some clothes until being assigned to duty were gone: "The days of the 'hard-boiled' methods in the training of recruits is passed." Drill instructors were still selected from the ranks of noncommissioned officers, some stellar graduated recruits or appointed "Acting Jacks." The latter were designated for various reasons, as exemplified by future Major General William P. Battell's selection in 1927. "I was the tallest man in the platoon."

The drill instructor's authority did not wane following the war. Although some officers occasionally drilled recruits and seemed active in recruit supervision, drill instructors remained "pretty much on their own." General Russell Jordahl recalled that when a company commander, he was advised to "'stay away from the recruits.' In other words, leave the DIs alone, you are there for administration only." Future Lieutenant General Edward W. Snedeker was a 1926 recruit company commander. The officer was informed that his job was "to go out and watch the drill instructors instructing the recruits, to keep my mouth shut and not to say anything, just to look wise, which I did very satisfactorily."

Parris Island

The physical requirements for recruits varied during the decade, with the lowest acceptable height ranging between five feet four inches and five feet six inches from 1925 through 1929. It appears that the recruit oyster shell details all but vanished in the 1920s, or became the exclusive work of the Parris Island naval prisoners for building base roads.

Recruits still came from the families of laborers, farmers and storekeepers, while others such as Stuart C. Stetson's father was the United States minister to Poland. Processed applicants were still offered the final opportunity of whether or not to enlist. The noncommissioned officers and the recruit manual made it perfectly clear to the applicant that the military did things a special way, but that hard work and attention to one's instructor would be rewarded in future years.

Applicants were usually sworn in at the receiving barracks, after which they were transferred to the recruit training areas. The Parris Island training grounds and barracks were known as the East and West Wings. Postwar enlistments and military reductions could delay a platoon forming for as long as a month. Once in training, and having received their rifles and other equipment and clothing, the recruits were placed under the drill instructor's care. The training period depended on any urgency for Marines and was normally eleven or twelve weeks. Some recruits received six or eight weeks training and others were in boot camp for as little as twenty-four days.

The initial training phase was at the main station area, mastering drill and being schooled in military subjects. The second phase was at the rifle range. In the third and final phase, additional or new instruction was offered, along with hours of drill. Other schooling included bayonet fighting, wall scaling and rope climbing. Additional instruction listed physical training, swimming, boxing and a classification processing of recruits. Upon the completion of training, or possibly during the cycle, police and mess duties were assigned.

It must be remembered that recruit training constantly changes, and that generalizations must be used. For example, the above curriculum varied at times. For instance, in 1928 the schedule included four weeks on the rifle range rather than three. With the onset of the 1930s Depression, the Parris Island Receiving Barracks (Quarantine Station) was deactivated, and rifle range quarters were closed as military budgets were cut. The deactivation of rifle range quarters necessitated the daily trucking of permanent range personnel from the base main station to the rifle range.

The Marine Corps prided itself on promoting officers through the ranks. Future lieutenant generals entering the Corps as enlisted recruits included James P. Berkeley and Thomas G. Ennis. Berkeley arrived at the depot in 1927, and received an indoctrination lecture that asked, "Do you still want to enlist in the Marine Corps?" About ten days were spent at the receiving

CHAPTER FOUR

area before the drill instructors met Berkeley's new platoon. The platoon then spent three weeks at the rifle range before returning to the main training area to complete boot camp. General Berkeley recalled there was no officer supervision and some hazing, but no maltreatment, which he maintained crept into the system in World War II.

General Ennis later commanded the depot. One of his most vivid memories was that in 1922, recruits were allowed only one bucket of fresh water per day. If the water was used unwisely and a recruit's supply was exhausted, salt water was substituted in its place.

Other enlisted men and officers never forgot their boot camp days. John Oxford reported to Parris Island in 1919, and remembered dropping to the ground repeatedly during imagined aid raid drills. At graduation, "the DIs just kind of told us we were done and were being shipped out." Other boots recalled Parris Island as a "tent city," also remembering the gas mask drills and wearing their gas masks on marches to the rifle range. If a recruit dropped his rifle he had to kiss or sleep with the weapon for one night or more.

Future Colonel Theodore M. Sheffield never forgot the gas mask drills, the heavy marching packs and that when his platoon graduated half of the new Marines were ordered to sea school and the others were sent to Nicaragua. Thousands of Marines departed Parris Island for places like Quantico or Charleston or to foreign assignments in the West Indies and China, as well as other places worldwide. Overseas departures were on occasion made from the Parris Island dock and on large ships that were especially utilized for transporting Marines. Old Corps Marines recalled the USS *Chaumont* and the USS *Henderson*, which anchored off of Parris Island.

Warren F. Lear was a gentle man whom the author met at the last Parris Island reunion of 1920s and 1930s recruits. Mr. Lear enlisted in Philadelphia in January 1920. The future officer's first order was to deliver himself and several other applicants to Parris Island following their Yemassee train ride. Lear requested the conductor to inform him when they arrived at Yemassee, which the conductor seemingly did. However, and to Lear's dismay, it was discovered the recruits were detrained 150 miles distance from Yemassee. Seeing his military career all but wrecked, Lear, with Marine ingenuity and alacrity borrowed a utility "hand pumper" to catch their train at one of its stops.

Lear's train ride from Yemassee to Beaufort and Port Royal was not without peril. The recruits were introduced into a coach filled with drunks, whose rowdiness was suffered until the train arrived at Port Royal. From there the group was ferried to Parris Island on a barge. A march along a swamp that was inhabited with "alligators and big snakes" brought the young men to the receiving barracks, where they were fed, bunked, and ordered by an instructor to go to sleep.

As the instructor turned the lights out, one applicant shouted, "Horse shit!" The annoyed Marine immediately reappeared. The recruits were ordered out of their bunks, instructed to place all of their clothing on the floor and to stand naked in front of their bunks. At this time the irate instructor turned a fire hose on the youths, who stood shivering in the cold Parris Island winter for the remainder of the night.

Private Lear's training lasted approximately three months, and included four weeks on the rifle range firing the Springfield '03 rifle, Browning Automatic Rifle (BAR) and pistol. Squad drill was taught. Since Lear had some previous military training, he was permitted to drill his recruit platoon. He promptly marched the platoon into a swamp. For this error Lear was ordered to carry 150 buckets of water from the washroom to his gunnery sergeant's office. It was an inauspicious beginning for Private Lear, who became an officer in the Dominican Republic Guardia Nationale and retired as a Marine Corps officer in 1954.

Future Major General Walter A. Churchill recalled Parris Island as being only accessible by boat and remembered that his clothing was washed in salt water. One's mess gear was cleaned daily. "The DIs would make us wear overcoats on hot days and our regular khakis on cold days" as punishment. Churchill's drill instructors were as loud and as vociferous as ever.

Private Theodore S. Talpalus was sworn in on the island in 1922, and assigned to a sixty-four-man company that was divided into two platoons of thirty-two men each. The company was marched from the receiving barracks to their assigned living barracks. Drill instructors also lived in recruits' barracks but had private rooms. (That practice was abolished in 1956.) Following physical exercises and cleaning the barracks each morning, drill and instruction followed in military subjects. Rifles, clothes and the recruit's metal mess gear were cleaned each night. If a mess kit did not pass inspection it had to be cleaned after taps.

"Monkey inspections" were examinations of a recruit's body, feet, hands and fingernails while the boot sat "bare naked" on his locker box. Any omission of cleanliness, or any other part of his training, could result in a recruit being ordered to run at double time around the parade ground or to run through a platoon belt line. If a boot insisted on looking at the ground during drill, an instructor or "Acting Jack" might give the culprit a short punch under the jaw. Talpalus found that such physical punishment was the exception rather than the rule.

Rifle instruction and firing was conducted after the recruit company marched to the range in the morning, and later returned for the noon meal

CHAPTER FOUR

in the main training area. Additional duties in 1922 included searching and digging for the historic French outpost of Charlesfort, the site marking the arrival of Parris Island's first European inhabitants in 1562. About fifteen recruits pulled ropes that were attached to buckets to remove any soil that buried historical artifacts.

Private George Watson's 1922 drill instructors welcomed new platoons to Parris Island with such remarks as "when the Corps gets to the point where it has to start taking such stupid egg-heads as you—then the Marine Corps is on its last legs." Along with the rifle training and drill, physical exercises were done on "stall bars." Other exercises included backbends, rope climbing, wall scaling and physical drill with rifles. The future sergeant major recalled that double-timing for a mile was done by the recruits who ran in their bathing suits.

Private Wilburn C. Creecy recorded his 1925 training in the East and West Wings, and the marches between their barracks and the rifle range. Medal of Honor winner Donald L. Truesdell was another 1925 Parris Island boot. Private Vincent Billings remembered his very short 1926 "Von Hindenberg" Marine Corps haircut and receiving dental care and fillings that lasted him all of his life.

Squad drill was practiced on the large sand parade ground, which was complete with prickly burrs and gnats. Body and clothing inspections were held on the parade field and in the barracks. A recruit's teeth were checked for proper brushing, and clothing displays were scheduled to see if a boot's apparel was properly washed. If the clothes did not pass inspection, drill instructors stomped them into the sand. Cigarette butts were meticulously hunted at police calls, as if they were pieces of gold. Rifles were field stripped on the parade field and the parts were memorized. Any angry disagreement among boots could be settled in the "bull ring."

Private Hoot Gibson enlisted in 1925, and was herded through the receiving barracks by Leland "Lou" Diamond. Lou was known for his distinctive voice and the noncommissioned officer achieved legendary fame during World War II. Diamond quickly reminded the young men that they were only applicants and not yet Marines. For Diamond anyone with less than ten years in the Corps was a boot. Gibson also remembered Captain Benny Fogg, who was the officer in charge of the receiving barracks. The officer was "a fatherly type man," but one who "would throw his boot at you without removing his foot if things were fouled up."

After about ten days a newly formed platoon was turned over to their drill instructors for clothing issue and personal items, including a bar of saltwater soap. Gibson stated "there was no fresh water on the island except for drinking, cooking and for the officers, and this was brought in by water barge." Only one fresh water spigot was in the east wing training area. It "was kept locked and guarded," except when recruits were issued one canteen of fresh water daily to drink, for shaving and for brushing one's teeth.

Each morning before sunrise Gibson's platoon marched on the parade ground that was "ankle deep in sand." Upon completion of their rifle firing, the recruits were assigned to work details such as cutting the rifle range grass or to work at the range officers' quarters. The platoon returned to their barracks prior to the evening meal to prepare for an inspection before supper.

Gibson had no regrets on leaving Parris Island, and he did not recall a drill instructor ever laying a hand on a recruit. Nevertheless, the DIs "all carried sabers and it was not unusual if you were out of step to feel the flat of said saber on your posterior." Problem recruits were singled out and matched in boxing and wrestling bouts.

Ralph T. Bowden entered the Corps as a 1926 private and recalled that the quartermaster "guessed our size and threw out our clothes." Recruits fired their rifles at two hundred yards standing, three hundred yards sitting, five hundred yards in the prone position and six hundred yards prone, with the rifles steadied by using a sand bag. Recruits under eighteen years of age were sent to the base field music school. They sometimes drew duty as range sentries, who manned a tower that was positioned in a Parris Island swamp at the rifle range. Whenever boats ventured into the impact zone a warning flag was hoisted and rifle firing was ceased until an intruding boat was removed.

Future Lieutenant Joseph E. Johnson arrived at Parris Island in 1926. At the receiving barracks, he was also met by Lou Diamond, who among other achievements became the Corps' most famous mortar man in World War II. Diamond informed the boots during one examination that "everybody will piss in this bottle tomorrow morning" (for a urinalysis), and one recruit had the temerity to ask, "From here?" Diamond replied, "No, you dumb son of a bitch," and Private Johnson began learning the ways of the Corps.

The normal three months training ensued for others when the recruits bounded out of their bunks at 0530, to the sound of the drill instructor's whistle. Once the platoon was ready to depart from its barracks, the DIs informed them that if anyone did not descend the stairwells (ladders) fast enough, they had better jump out of the windows and be at attention when a DI fronted the platoon—even if they had window frames around their necks. The drill of running in and out of the barracks and up and down the narrow stairs was most likely common in the twenties, a practice well established in the 1950s. Yet Johnson realized that hazing was the price of

CHAPTER FOUR

becoming a Marine and he never resented his DIs. "They were firm, they were a little bit hard; but nobody was beaten up and all this."

Major Robert A. Smith was attracted to the Corps by a recruiting poster that "depicted a husky, bronzed Marine dressed in khaki, a campaign hat cocked jauntily on his head." After being barged to the island in 1928, Smith's group was taken to a small wooden barracks that was heated by two potbellied stoves. Blankets, linens and pajamas were issued, and instructions were given on how to make one's bed or "bunk." The boots were later fed supper and treated to view a film entitled *Tell It to the Marines*.

After being confined to the hospital for ten days with a fever, Smith met his three drill instructors, who informed the platoon that all DIs would be addressed as "Sir." If anyone doubted that a drill instructor was not "tough enough to lick any man in this outfit," recruits were invited to step out. No one did. Shortly, clothes were issued, and haircuts, shots and vaccinations were administered. Boots also received their bucket issue that consisted of personal items.

The training area barracks were double-decked wooden structures that were heated by coal-burning stoves. Smith's training consisted of the standard instruction of drill and classes plus physical exercise. Any recruit who failed to finish double-timing around the area could be punished for falling out. One punishment was pushing a loaded wheelbarrow through the Parris Island sand. The belt line was also used. Sometimes DIs ordered recruits to run through the gauntlet for "no reason at all." DIs dissatisfied with the condition and cleanliness of a barracks would wreck the squad bay or throw items out of the windows to harass recruits.

Much of the training made little or no sense to many other than instilling the lesson of obeying orders with no reservations whatsoever. For example, several "short arm" venereal disease inspections were administered during Robert Smith's training, even though the recruits had not been near a woman in weeks. Just prior to Smith's Parris Island departure, a sixty-year-old Navy nurse performed the final "short arm" inspection.

The recruit depot was minus many vestiges of World War I by 1930, as Parris Island slowly became more accessible to the mainland. A narrow causeway and a two-lane bridge were dedicated in 1929, and aircraft and automobiles were seen on Parris Island by 1930. However, with the 1929 financial crash on Wall Street, Parris Island's expansion and construction all but ceased until the eve of World War II.

The Yemassee railroad junction was the South Carolina destination for Parris Island bound recruits until 1965.

CHAPTER FOUR

PARRIS ISLAND

A recruit's railroad journey from Yemassee to Port Royal could be in a coach with veterans, drunks or a conductor who enjoyed teasing the civilians who were about to begin their training.

A trained platoon of World War II recruits marching by new arrivals. The leg display could be one of the seemingly insane drills of former boot camp days.

CHAPTER FOUR

Arriving recruits were once received by drill instructors at the base Headquarters & Service Battalion Building.

The two speeds at Parris Island are fast and faster. Newly arrived recruits exiting their bus.

Since the early 1960s, the Parris Island journey began from the famous yellow footprints.

Newly arrived recruits "get the word."

Chapter Five

WHEN A BOOT LEAVES PARRIS ISLAND, HE TALKS LIKE A SAILOR, WALKS LIKE A SOLDIER, BUT BOASTS THAT HE IS A MARINE.

KATHERINE M. JONES

The Great Depression severely curtailed military expenditures, and it was speculated that the Parris Island base might be closed. Although the battle to save the station was won, there were casualties of both material and men. Some base beautification continued as the Great War became a memory. *Leatherneck* magazine reported in 1931: "You old timers who knew the Island during the big argument across the pond would be surprised to see it now, for it is no longer the barren spot it used to be. Today it is a sylvan retreat of the South, with trees, green lawns and fields of the kind that constantly remind you of nature's handiwork."

Yemassee grew during the Depression as the town was adjacent to two highways and received trains hourly. The Greyhound bus company operated between Beaufort and Yemassee, as the railroad junction remained the last stop for civilians who were about to become boots.

The training time and the Parris Island recruit population varied widely. The minimum numbers seems to have been during the depths of the Depression, with only 500 recruits processed in 1932. Some 750 men were processed in 1933. One 1935 platoon began training with as few as 8 recruits, and by January 1936, there were seldom more than two platoons on a training schedule at one time. Platoon 8 had 18 recruits and three drill instructors in 1938.

The erratic recruit numbers frequently caused drill instructor shortages, which were alleviated by recruit "Acting Jacks," or the appointment of privates in the 1930s as acting sergeants and corporals. These acting promotions were payless, and acting ranks only had authority over other boots. Such temporary NCOs were sometimes identified by wearing only one of a set of chevrons on one sleeve. Yet *Leatherneck* magazine carried photographs of recruit platoons of the 1930s that showed DI privates wearing no chevrons but with acting ranks. All permanent DIs were identified by a full set of chevrons and a swagger stick to be used to administer a "smart tap on the butt to let you know you had goofed."

Whether the instructors were sergeants or appointed recruits, all in authority applied a code of strict discipline in the process of turning civilians into Marines. Individuality was torn away by referring to recent civilians as "dummies," "birds," "bozo," "burr-headed buzzards" and the popular World War II "Joe," which was used as early as 1933. Recruits were ordered to "Take that cigarette out of your face, Joe! Take your hat off! Stand at attention! You're in the Marines now!" Other commands included, "Y' gotta snap outta yer dope from now on," and any boot that doubted an instructor's sincerity could be invited to step forward. Few did.

On other occasions recruits were told to "Keep your chins up, or I'll lift them for you." A swift kick was another threat. Requiring errant recruits to roll a marching pack underneath a bunk was ordered, as was standing at attention while annoying sand fleas crawled over a recruit's face, into his eyes, ears and in his hair. Under such circumstances a DI screamed the worst threats if a recruit so much as flinched.

Repeated to-the-rear marches in the sand and choking dust was part of the training. If a boot fell from exhaustion, the platoon might be ordered to march over the prostrate recruit. A last ounce of strength, like that which saves lives in combat was found, or the DI would halt his platoon before a trouncing occurred. Recruits who received postal letters addressed "Mr." rather than "Private" were sentenced to run the belt line. The drill instructor's authority was clearly evident and omnipotent, as illustrated in a humorous account published by *Leatherneck* magazine in 1935.

> *One day a recruit who had not been on the island very long was hurrying to join his outfit and happened to pass an officer whom he failed to notice and salute. The officer called him back and gave him a few instructive remarks about military courtesy. The recruit listened with an anxious eye toward his waiting platoon and the instructor in charge of it. When the officer finished speaking, the recruit turned and hurried away, again without saluting. Again the officer called him back. The recruit was in a quandary. "You'd better let me go now," he said, "or the sergeant will come over and give us both hell!"*

Recruits have perennially viewed drill instructors' methods as "endless, unnecessary and disjointed." The recruit's curriculum was seldom appreciated until the training was complete. First Lieutenant Wallace M.

CHAPTER FIVE

Greene Jr. turned his attention to recruit training in 1936, realizing that "a tree is no better than the soil which nourishes it." The future Marine Corps commandant stated that not enough attention was given to recruit training, and that recruiters at times filled their quotas by enlisting undesirable men. Greene believed that boot training should be extended to a minimum of eight weeks, and that particular emphasis should be placed on the obedience and confidence that was achieved through close order drill. The officer also championed the need for a Marine Corps manual for recruits. The first *Marine's Handbook* was published in 1938.

Lieutenant Greene ranked the importance of conditioning near that of close order drill, even though the "wails from the dispensary" and "letters from mothers who worried over Joe Recruit's black eye or sprained ankle" clearly outweighed the fact that "Joe is learning to take care of himself." Greene also realized the need for a permanent drill instructor's school, to train men for the most "vital task" in the Marine Corps in peace time or war.

Many of General Greene's recommendations were later adopted. However, in 1931, former Marines were questioning some training changes that were taking place. *Leatherneck* magazine noted in July 1931, that recruits neither trained all hours of the day and night nor were required to endure saltwater showers in "freezing blasts of wind." Hikes to the range or other places the drill sergeants decided on were a thing of the past. The magazine reported that trucks were now used to transport recruits, and that drill was "done on regular schedule and no other is allowed." It was added, "Soon we will be giving them [the boots] breakfast in bed."

Perhaps at no other time did the Marines enjoy such a choice selection of qualified recruits. Depression enlistment requirements were exceptionally high, and recruiting duty was described as a "cinch." Given this advantageous situation as early as 1931, the average applicant was practically eliminated from becoming a Marine. Applicants had to possess a high school diploma to even rate an interview. If one passed their examinations and qualified otherwise, he was sent home to wait his turn to enlist.

During the early and mid-decade, recruits were required to have a minimum height of five feet eight inches, while the maximum age of thirty-five years was lowered to twenty-five years for veterans. Major General Oscar F. Peatross recalled that during much of the decade one could not "get in the Marine Corps unless you were a high school graduate." An applicant had to have near perfect teeth and was placed on a waiting list as part of a quota before being admitted into the Corps. The general stated, "There's no question but what the Marine Corps had a pretty high caliber of man during the thirties" and until the eve of World War II.

Parris Island

As that war seemed irrepressible, the Corps' height requirements were lowered in 1937, and applicants could enlist with only twenty-nine healthy teeth. Moreover, President Roosevelt's alerting of the military services to the possibility of a war placed a strain on the island, causing the construction of huts, tents and Personnel Barracks (PB) that were built with warped green lumber. The shabby buildings became home to thousands of World War II Marine recruits.

By 1940 there was a drill instructor shortage and recruit training was reduced to only four weeks, including the rifle range. Recruit arrivals jumped from about 190 men per month to as many as 730 recruits reporting to the depot in one day by late 1939.

Such a large influx of recruits and the reduced training schedule caused serious concern. For example, the one week devoted to marksmanship saw the percentage of recruit rifle qualification drop from as much as 48 to 25 percent at Parris Island and from 92 to 67 percent at San Diego. The more favorable San Diego figure was attributed to having more trained rifle coaches and superior weather in which to shoot.

Marine Corps boot camp remained a routine of discipline, drill, physical exercises and marksmanship instruction of varying durations up to World War II. Measles continued to plague unfortunate recruits, and one 1934 platoon was placed in quarantine in a former prison brig for twenty-one days. Cleanliness was always stressed. Boots welcomed the demise of the old wooden wash racks, which were replaced by "long smooth concrete tables with convenient faucets and excellent drainage facilities." A few of these concrete racks survive at Parris Island, and can be used for current recruits to wash their field training gear.

A 1931 base pamphlet informed recruits that they could expect to be clothed, fed and put to bed on their first day on base. The second day was addressed to taking exams, receiving haircuts and being sworn in. With an additional issue of clothing, equipment and a rifle the recruit was assigned to a sixty-six-man platoon and sent to his training area.

Basic military subjects were taught during the first three weeks, including bayonet and constant drill. The fourth week was devoted to "snapping in," or practicing shooting positions for hours, while aiming at large black dots painted on white aiming posts. The fifth and sixth weeks were devoted to the live firing of the 1903 Springfield rifle, which was capped by the formal and single rifle qualification day.

The 1930s range course required proficiency at the two-, three-, five- and six-hundred-yard lines. Slow and rapid fire was tested, and a shooter had to master the standing, kneeling, sitting and prone positions. Marksmen were required to shoot 275 points out of 350, sharpshooters had to score

CHAPTER FIVE

300 points and a rifle expert had to record 315 points. Additional pay was awarded to superior shooters. Some 1930s platoons qualified as many shooters as 95 percent. Recruits also received a familiarization firing of the Browning Automatic Rifle (BAR) and the Thompson sub-machine gun and attended demonstrations of the trench mortar. Drill instructors may have served as rifle instructors at times.

The final two training weeks included more drill, schooling and participation in a physical training program that was recalled as "all work and no play."

The Depression recruit often found a home in the Marine Corps. Others were attracted by the Corps' reputation of being "The First To Fight," commercial Marine movies, the dress blue uniform and cocky "Devil Dogs" or "Leathernecks" wearing their "salty" campaign or field hats. But any romanticism was immediately dispelled once one arrived at Yemassee junction and the Parris Island dock.

Major William E. Hemingway began his military career in 1930, and was assigned to a barracks infested with "millions of bed bugs." His lengthy fourteen weeks in boot camp were devoted to eight-man squad drill and the live firing of the rifle, pistol and the BAR. Roy E. Corbin in 1930 remembered his Marine haircut, the heat, sand fleas and drill instructors who shepherded a thirty-two-man platoon for three months.

Private Walter Wilson crossed from Port Royal to the depot by motor launch in 1931. Wilson lived in a wooden barracks that was "next to a marsh." His platoon was composed of about sixty members that trained for twelve weeks. The recruits were schooled in basic military subjects by three instructors, and there were always the mosquitoes, flies and sand fleas in fields of sharp sand burrs.

Private Joseph Hornstein was a 1930 "Acting Jack," who recalled that only one platoon on the island could carry a flag designating it as the honor platoon. The future warrant officer also remembered that the DI selection was not as meticulous then as it later was. For Private Carl E. Hardy, the base was pretty much like Devil's Island. The future sergeant major offered, "You had the impression that Parris Island was rather isolated and that you were at the end of the world."

Two other Marines who began their careers at Parris Island in 1931 and 1932 were Commissioned Warrant Officer George T. Edwards and Lieutenant General Lewis J. Fields. Edwards was temporarily assigned to police duties, until a sixty-six-man platoon was formed for a nine-week training period. Like others before him, hours of sweat and effort were spent on the parade field, which was little more than a huge sand lot with barracks on the bay or water side. Upon his completion of boot camp

Edwards was assigned to a depot "War College." The "College" may have been the noncommissioned officers' school, where some DIs were trained as NCOs.

General Fields's recruit platoon contained a number of high school graduates and college men from Notre Dame and West Point. Yet some of these college-educated men were not clean enough for the Marines. One recruit paid little attention to his uniforms or to his hygiene. His platoon scrubbed down the errant boot with strong Octagon soap and scrub brushes, that left some redness and little streaks of blood on the thoroughly cleaned recruit. The lesson was effective and the neglect was not repeated again.

The Medal of Honor was awarded to Private Mitchell Paige on Guadalcanal. Paige recalled that he "never realized anybody or anything could be so rough" as our drill instructors, who "were truly tough men." The future officer recalled working in a mess hall, "swabbing, scrubbing pots and pans and waiting on tables, from 0400 to late at night." Other arduous training was at the bayonet course, which had a sign reminding recruits that

A BAYONET FIGHTER KILLS OR GETS KILLED!

Paige's drill instructors on another occasion marched the platoon into a rifle range tidal stream. "The first man in the water marched straight through, the rest followed, and as a consequence nobody was hurt." Years later, Colonel Paige offered that "six Marines drowned when marched into that same creek [according to Paige] by their drill instructor, but apparently they did not maintain march discipline." Graduation ceremonies were "the highlight in my life to that point." Paige's platoon paraded behind a Marine band, and he saw a Parris Island officer for the first time.

Private Henry T. Dawes was quartered in the main station recruit barracks in 1937. At sometime he was impressed seeing World Heavyweight Champion Gene Tunney's name carved on one of the old barracks. The recruits threw practice hand grenades and fired the water-cooled machine gun during weapons training. Toward the end of Dawes's boot experience, the platoon was quarantined for mumps. The recruit retired as a lieutenant colonel, and recalled that his DIs were fair and "straight shooters," who taught the platoon eight man squad drill. Yet there was always swift punishment for the boot that was out of line.

Carter Fisher journeyed by railroad from Philadelphia to Yemassee. Recruits never forgot the small South Carolina town that was the subject of recruit humor. For example, in the 1930s boots had to recite, "I'm a horse's ass from Yemassee because I lost my locker key!" There were other

CHAPTER FIVE

variations. A bucket issue was assigned to recruits, and if a boot did not do as instructed, one may receive "a kick in the butt." Facial slaps from the ground up were known.

The recruit wooden barracks were partly in a swampy area and duckboard walkways spanned the dampened ground. Each building had front and back steps, and one could see the sky through cracks in the barracks roof. Heads (bathrooms) were either in a shed or on the first floor.

A platoon then consisted of six squads of eight recruits. DIs used whistles for communications and orders, and there were three DIs assigned to Fisher's platoon. One corporal was recalled as being sadistic. Hazing seemed commonplace and disaster followed the recruit who received a letter addressed "Mr." at mail call. Private Fisher's graduation was seemingly routine since "we just finished and were dispersed worldwide at $20.80 per month."

Private Samuel Cosman was a 1945 drill instructor, who remembered his 1938 recruit training as "tough." There was a lot of "troop and stomp," and the omnipresent Parris Island sand flea attacked recruits while the boots stood at attention. Each recruit had his own clandestine sand flea resistance to escape the watchful eyes of the DI. Cosman formed his lips "in such a way I could blow thru my lips and have the air blow the sand fleas away." Drill instructors referred to their recruits as "horses' asses from Yemassee," and a recruit who dropped his rifle had to sleep with it or with as many as nine more. Running in and out of the barracks with heavy wooden locker boxes and physical drill underarms was practiced, while on other instances recruits were ordered to double time around the sand drill field with their Springfield '03 rifles carried at high port.

Private Cosman recorded that campaign hats were the standard issue in the late 1930s, and the recruit work uniform was blue bib-type overalls that were supported by suspenders. A private's pay was $21.90 a month. Funds were subtracted from one's boot camp salary in payment for a bucket issue of toilet and other personal items. One was a straight razor, and a recruit had to shave whether one needed to or not. There was no free life insurance at that time, but other policies were available at a cost of about $3.20 a month.

Clothing issues were as matter-of-fact as the instruction of drill. When the recruits were issued their field shoes, called "boondockers," they held a five-pound dumbbell in each hand that "was supposed to add weight to your body which applied pressure down on your feet." This possibly assimilated the wearing of a heavy marching pack.

Thefts and clothing raids by other platoons were always a concern. One safeguard was the use of clotheslines that were over ten feet high. A line near Cosman's barracks would be lowered to the ground, the clothes would be hung, and the line would be elevated and suspended between two large poles. The mechanism that hoisted the clothes was locked in a box. One trustworthy recruit was assigned to be responsible for the key.

Private Cosman recalled that reveille every morning except on Sunday was the duty of a depot drum and bugle corps that marched down the main street. Recruits refought the Civil War, and farm boys seemed to be easy targets for ridicule. One joke was, "They had to put plow handles on the train to get you to Parris Island."

Rifle qualification "was our number one objective." Recruits who failed to qualify were informed that their enlistments would be spent on mess duty, and shooters with scores under par had their performances read aloud. The pressure to qualify was enormous since boots were taught that a Marine's rifle was his right arm.

Private Wilbur H. Stuckey's day began when the DIs whistle announced the training day at 0500. If a platoon did not spring from its bunks, or exit the barracks to the drill instructor's approval, practice bunk drills were the first order of the day. On one instance Stuckey was slow and was threatened by a DI. "I am going to have you moving so fast that when you turn a corner you will blow sand in your shoes."

The DIs doubted if a platoon could ever please them, "but that you are going to damned well try." Besides marching in the sand and offering their bodies to the insects, Stuckey recalled one unfortunate recruit who locked his locker box key inside of the large heavy wooden box. The boot reluctantly approached the DI who had the matching key. The drill instructor offered to unlock the box, but the recruit had to carry it up a flight of stairs to the DI room, and then back to the first floor. From that day on the platoon was attentive to the whereabouts of keys, and to the fact that, as in combat, each and every error required a price be paid. In later times, one locker box key was tied around recruit's necks and drill instructors retained the duplicate key.

In another instance two recruits concocted a plot of feigning exhaustion. One boot fell to the ground and another would stop to rest while pretending to offer aid. Corporal Scott had trained his fair share of platoons, and he quickly detected the ploy. The drill instructor immediately grabbed the prone recruit by the collar, screaming, "Damn you, if you fall out in my platoon you better damn well turn pale." Additional laps were awarded the two recruits and the Marine Corps lesson of never quitting was learned.

Stuckey recalled in 1939 that it was "not uncommon to be struck by a DI when we were in training, although I never knew personally of anything worse than a face slap or a sharp nudge in the gut." Private Stuckey was

CHAPTER FIVE

once struck in the stomach with an elbow and ordered to "suck in that gut!" Such treatment hardly made drill instructors objects to be loved. But in later years they were regarded as "masters at the art of drilling and using salty language," who had jobs to perform. The consensus of Stuckey's platoon was that the DIs were good Marines who carried out their responsibility of training recruits, "conscientiously and well."

Although Private Stuckey never recalled "anything worse than a face slap or a sharp nudge in the gut," Private Bill Gulley's drill instructor appeared to be a more violent man. Gulley enlisted in 1939 at the age of sixteen, and recalled that it was in boot camp where he first learned "the meaning of fear." When Gulley knocked over a cluster of stacked rifles, a drill instructor "brought his fist up, and broke my nose." The private was told to see a corpsman (medic), "but whatever you do, don't you tell him how it happened. Not if you want to get out of here alive." Gulley was frightened, "and it was my first exposure to real abuse of power."

Parris Island survived the terrible Depression and began training recruits by 1940 for another World War. Recruits, who were destined to storm Second World War Pacific beaches, valued the rigors of boot camp and their rugged DIs as did the "Devil Dogs" of World War I.

Chapter Six

TO THE PACIFIC'S DISTANT SHORES, 1940-45. THIS, FELLOW, IS A DIFFERENT WORLD THAN THE ONE ALL OF US CALL HOME. THIS IS A MARINE WORLD—PARRIS ISLAND—NOT A SHANGRI LA, BUT A MELTING POT FROM WHICH IS WELDED THE STRONGEST LINK TO FREEDOM AND HOPE—THE UNITED STATES MARINE...
PARRIS ISLAND BOOT 1944

Following the December 7, 1941 Japanese attack on Pearl Harbor, the United States immediately entered World War II. San Diego's and Parris Island's populations realized their greatest numbers since World War I. More recruits arrived in thirty days than reported to the two depots in past single years. For example, at least 5,276 recruits arrived on Parris Island from December 7 to December 31, and five new recruit battalions were activated. Recruits were housed in barracks, huts and tents. The arrival of 9,206 recruits in January 1942 necessitated the formation of the Ninth, Tenth, Eleventh and Twelfth Recruit Battalions by midmonth. Such an explosive expansion caused equipment shortages and placed tremendous burdens on the Marines, who worked six and a half days per week for eight to twelve hours per day. The sluggish Depression years were no more.

The Parris Island recruit load neared 13,000 by August 1942. That required the organization of a record-making Thirteenth Recruit Training Battalion, and the creation of "New River Battalions" (Camp Lejeune) to assist processing the deluge of recruits. The New River units were dispatched to North Carolina for rifle instruction and for the completion of their course as basic Marines.

Parris Island was inhabited by over 18,000 recruits when the Pacific war ended in 1945. Hundreds of regular Marines awaited discharge, as base separation centers were created to handle the large numbers of men. Other offices were levying drafts for China-bound Marines, as depot personnel vacancies were frequently filled by graduating boots.

The Corps' enlistment age was lowered from eighteen to seventeen years, in 1941 with enlistments for the duration of the war. Selective Service was used also. The course-training schedule varied as it had in 1917-18. For

CHAPTER SIX

example, in 1939, boot training was halved from eight to four weeks. In February 1940, the training period was extended to seven weeks. Recruit training was not lengthened to eight weeks (which included three rifle range weeks) until 1944. The cycle remained at eight weeks until the end of the war in September 1945.

Leatherneck magazine in March 1945 described the "Drillus Instructor" as a "horn-headed bug" that preys on recruits. "Violent mental anguish and spasmodic physical contortions accompany victims of this bug from eight to twenty weeks." For certain, the World War II drill instructor demanded discipline, pride and the dedication of recruits to the Marine Corps. Some boots feared their DIs more than the Japanese. However, it is difficult for one who has not experienced Marine recruit training to understand that this fear was a positive, or a "good" fear to be appreciated on a battlefield. Some recruits even "loved" their drill instructors.

Drill instructor training remained almost entirely on the job. The instructors shocked and stripped their recruits of any civilian identification, and the boots were constantly kept off balance in a stressful environment. Paradoxically, DIs could be compassionate, sarcastic and cruel. Recruits were informed, "Nothing is too good for you, but we'll let you have it anyway."

The DI was a master of profanity who could make the proverbial sailor blush. Recruit orders were generously punctuated by that vile four-letter word that identifies the act of sexual intercourse. Former recruit Robert Leckie skillfully wrote in his book *Helmet for My Pillow*:

> *Always there was that four-letter ugly sound that men in uniform have expanded into the single substance of the linguistic world. It was a handle, a hyphen, a hyperbole; verb, noun, modifier; yes, even conjunction. It described food, fatigue, metaphysics. It stood for everything and meant nothing; an insulting word, it was never used to insult; crudely descriptive of the sexual act, it was never used to describe it; base, it meant the best; ugly, it modified beauty; it was the name and the nomenclature of the voice of emptiness, but one heard it from chaplains and captains, from Pfc's and Ph.D.'s—until, finally, one could only surmise that if a visitor unacquainted with English were to overhear our conversations he would, in the way of the Higher Criticism, demonstrate by measurement and numerical incidence that this little word must assuredly be the thing for which we were fighting.*

Profanity was common in the other services, too. In 1941 the U.S. Army at Fort Devens, Massachusetts, attempted to forbid any profanity while training recruits. Many saw profane language as an abuse. Others mistakenly argue that profanity demonstrates one's poor vocabulary.

Throughout the lives of most Marines, the DI and his inimical "Awn awp reep…reep fawya laf," was indelibly etched in their minds. Mr. Leckie never forgot his DI or the marching cadence. Leckie remembered that Sergeant Bellow was a six-foot-four-inch, 230-pound, semi-illiterate Southerner who was a master of that peculiar Marine Corps "lilting cadence of command." "Thrip-faw-ya-leahft, thrip-faw-ya-leahft" was no more than "Three-four-your-left," Marine Corps style. Other orders included, "Right shouldeh, ahms," followed by "Strike your pieces, y'hear? Ah want noise! Ah want blood! Noise! Blood! Present, ahms," and numerous other commands that the recruits mastered through the obedience that comes with unrelenting drill.

On other occasions the sergeants bemoaned that they had the most stupid recruits on the entire island, and that God and the Marine Corps had cruelly sentenced them to train such "shitbirds," as if some divine scheme of retribution was being inflicted on the NCOs. Yet Leckie recalled, "on the whole, the sergeants were not cruel. They were not sadists. They believed in making it tough on us, but they believed this for the purpose of making us turn out tough."

Other former recruits and prominent authors wrote of their DIs. Gilbert P. Bailey recalled that his drill instructors were teachers, taskmasters and the key men "in the boot camp process." They were the "living example" of the word "Marine." Author Jim Lucas remembered Parris Island as "an unbearable burden" that could be tolerated only by fomenting a "burning hatred" of the DIs.

T. Grady Gallant wrote that "Blaskewitz was a good DI," but that recruits' mothers would not have liked him very well:

> *He represented all the Marine Corps instructors in the world in our eyes. We were his flock; the sheep of his sandy pasture. He protected us from outside forces. He was law. He was the judge and jury, the prosecutor, the high court of appeals. His word was final. His instruction was incontestable. His wisdom was supreme. He was authority, as we knew it. He was Justice raining on the deserving and undeserving such blows as he felt the occasion demanded. But he knew the secret of firm, demanding leadership; a leadership that applied pressure to a certain level—to a certain critical point—then relaxed ever so slightly to prevent breakage of the delicate tower of good he was slowly building.*

The Marine Corps has never officially sanctioned hazing and maltreatment, nor did the former "Rocks And Shoals" and the current "Uniform Code of Military Justice" (UCMJ). Furthermore, only a small amount of published

CHAPTER SIX

material existed by the mid-1950s that freely discussed the harsher methods used by past DIs. Previous physical instruction was simply accepted as part of becoming a Marine—something the uninitiated may not understand. Some argue that World War II DIs began excessively punishing recruits after the Marines returned from the Pacific War, so as to teach the Marines in the making the stress that combat can require.

Another argument offered to explain maltreatment in recruit training is that when the Korean War Reserve Marines were mobilized, they contributed inexperienced DIs into the system that resorted to "thumping" and brutality. Both arguments are weak. As the previous chapters have shown, hazing and maltreatment have been evident at Parris Island since 1917. It is unjustified to place any blame of the troublesome period following the Korean War on the shoulders of the World War II and postwar DIs.

Marines have always prided themselves with their mastery of the rifle, which is in the Marine Corps tantamount to discipline and esprit de corps. From the moment the boot received his rifle, he learned that it must never be abused, and that dropping the "piece" resulted in his taking the weapon to bed that night or for as long as a week. The practice was long honored.

The rifle's cleanliness was as important as the bodily hygiene of a boot. Throughout a recruit's training a Marine-to-be was harangued with a series of life-saving what ifs? "Private, what if you were in combat and your weapon failed to fire!" "Joe, neglecting the care of your rifle could cause the other men in your squad to be killed!" The lessons learned were that combat Marines depended on one another, and that your rifle was your close and best friend.

The sanctity of the rifle was imbued into recruit minds by the "My Rifle" creed, to first appear in *Leatherneck* magazine in 1942. The creed was written by Major General William H. Rupertus, and was printed in Marine Corps rifle scorebooks.

My Rifle

This is my rifle.
 There are many like it, but, this one is mine.
 My rifle is my best friend. It is my life. I must master it as I master my life.
 My rifle, without me is useless. Without my rifle, I am useless. I must fire my rifle true. I must shoot straighter than my enemy who is trying to kill me. I must shoot him before he shoots me. I will...
 My rifle and myself know that what counts in this war is not the rounds

we fire, the noise of our burst, nor the smoke we make. We know that it is the hits that count. We will hit…

My rifle is human, even as I, because it is my life. Thus, I will learn it as a brother. I will learn its weaknesses, its strength, its parts, its accessories, its sights, and its barrel. I will keep my rifle clean and ready, even as I am clean and ready. We will become part of each other. We will…

Before God I swear this creed. My rifle and myself are the defenders of my country. We are the masters of our enemy. We are the saviors of my life.

So be it, until victory is America's and there is no enemy, but peace!

One of many boot camp memories was squeezing off that first live round, feeling the rifle kick into the shoulder pad of the shooting jacket and awaiting the score result. After each round, or following a string of rapid fire, the targets were lowered on chain-operated carriages by other recruits, who were protected by earth and concrete rifle range butts. The danger was minimal, but now and then a target puller could be showered by dirt from a low shot hitting the top of the butts.

Recruits pulling the targets were urged on as if they were galley slaves. Loudspeakers ordered, "GET THOSE TARGETS OUT. GET 'EM IN. CAN'T YOU MOVE? GET 'EM UP! WHAM BAM, LITTLE MAN. GETEMIN…GETEMIN…GETEMIN. MARK 35. GOD DAM'T MARK 35! PASTE 'EM UP. RUN UP A-A-ALL TARGETS."

A pause followed as the line of recruits readied to fire again. "Lock and load" was the command to put the rifle on safe and to insert the live rounds. Once all safety precautions were checked, the range officer commanded, "Ready on the right…Ready on the left…Ready on the firing line. Unlock…TARGETS!"

A rifle designed for the sole purpose of killing one's enemy dug into the shoulder as a string of rapid fire was expended. The air was punctuated by the strong odor of powder, the ejection of spent cartridges and the metallic ring of the M1's clip as it was regurgitated from the rifle chamber after spending the eighth and final round. After all of the shooters fired the course and policed their "brass," the shooters marched to the butts to become target crews for those who served them.

Any failure to qualify with the rifle was once a near disgrace and could contribute to a recruit's suicide. Nonqualifiers were often marched at a rear distance from the platoon, and forced to wear their rifles with slings around the neck. Being called "stupid" and not having what it takes was also part of the disgrace, in addition to not receiving a marksmanship badge upon the completion of recruit training. Second chances were not offered to qualify.

CHAPTER SIX

Such former practices may be regarded as cruel in present times. However, the lesson regarding the significance of mastering the rifle, and not letting one's self or his unit down was learned again.

World War II and later Marines were grabbed by their collars, pushed around, run in and out of barracks and up and down stairwells (ladders). DIs placed buckets on recruit's heads, which was pictured in *Leatherneck* magazine for the Marine Corps hierarchy and the public to see. The bucket drill was also filmed in the 1955 movie *Battle Cry*. The penalty for putting one's hands in one's pockets was to fill the pockets with sand and sew them shut. Drilling for several hours in sand-burdened trousers insured that such an unmilitary practice would not happen again. Recruits, or even part of a platoon that might sneak into a post exchange to purchase ice cream, were required by some DIs to wear the carton on their heads, while the ice cream melted in the sun. Physical conditioning with rifles was done for exercise or punishment, and offenders who called their rifle a "gun" had to repeatedly recite verses such as:

> *I'm a yard bird from Yemassee,*
> *I called my piece a gun, you see,*
> *So woe is me*
> *Wh-hoe is me.*

Holding the rifle in one hand and grabbing the crotch with the other, recruits recited:

> *This is my rifle,*
> *This is my gun.*
> *This is for fighting,*
> *And this is for fun.*

It is reported that some DIs emotionally broke "problem" recruits and slow learners to rid them from a platoon. But there is no question that during the historical zenith of the Marine Corps, and later, maltreatment was a part of the boot camp process.

Any resemblance to being a civilian was altered at the hygienic unit. Recruits received their initial Marine Corps haircut and were subjected to physical examinations and clothing issues that left them in shock. One expression barked at the cramped lines of nude men was, "Close it up until the man in front of you smiles!" The humiliation was intended to reduce all recruits to the most basic and common denominator. The process was yet another part of becoming a Marine.

Parris Island

Master Gunnery Sergeant "Lou" Diamond served at Parris Island several times, including his tour in the hygienic unit in the latter part of World War II. "The Honker" had previously lectured arriving recruits until a Navy officer reprimanded Diamond for cursing at recruits. Not one to discipline his tongue, Lou is said to have told the chaplain that he should take care of the boots on Sunday, and "I'll take care of them the rest of the time." Soon, Lou and his dog were transferred to the hygienic unit (the "delousing plant") where Diamond's vocabulary was more in style.

The recruit's dress changed during the war as the Corps adopted a herringbone material for its field uniform. Several styles had large leg and rear pockets, or pockets that were the regular size. The same was true for the utility coat that had metal buttons inscribed "U.S. Marine Corps." When a recruit's utility coat was not tucked in the trousers, the boot resembled a prisoner in appearance. This too was another aspect of transforming a civilian into a Marine.

A female recruit training battalion was established on Parris Island in January 1944. The need was the same as for the distaff World War I Marines—to free a man to fight. The female recruits did not train as the men did, but drill and a regimented life was part of becoming a female Marine. Female recruits were trained by male drill instructors, with female noncommissioned officers present.

Recruits learned to live and survive among total strangers in the barracks, huts and tents. The Civil War was refought, fights occurred and there were thieves. If a piece of government equipment was missing, the recruit had to replace it. Boots either learned to be quick of hand in mess halls or to be satisfied with what food, if any, was left on one's metal tray. Southerners missed their grits and Marine green–looking liver was enough to make one gag. Recruits were also introduced to creamed beef on toast, fondly remembered by veterans as "SOS" ("Shit on a Shingle"). Yet on the whole, the food was good and recruits lost or gained welcomed weight and developed their muscles in boot camp.

Whether the recruit was a slow learner or an officer-to-be, he was subjected to the harassment, humiliation, sweat, cursing, shock, fear, homesickness and complete surrender to authority; that was part of becoming a Marine. Most boots had little difficulty dealing with their ordeal, and there was the satisfaction that came with the knowledge that one had challenged the arduous system and won. In the process, many Marines acquired experiences that served them during the war and in civilian endeavors in later years. The testimonials are numerous.

Pacific War veterans acquired the hindsight to know the value of the demanding training and gave lectures and wrote articles to communicate

CHAPTER SIX

their knowledge to recruits in training during the war. Captain Charles A. Vassey was wounded at Eniwetok and claimed that wars were "won in boot camp." The officer stated that boots in training think that a great "amount of the program taught them is nonsense—but let me say that you can't learn too much." Vassey added, "Every phase of boot camp serves its purpose in combat and there is no substitute for a Marine rifleman with a bayonet." During his time as a recruit, "the fighting man gets the start that will make him a poor, average, or superior fighter later on." Other veterans concurred and argued that Parris Island and San Diego were where the "real Marine is made."

Like his father in World War I, William Manchester Jr. was a Parris Island recruit. The future and famous author was aware of Parris Island's reputation, which "was marginally better than that of Alcatraz and Devil's Island." Once in training Manchester learned Corps jargon and such words as "slopchute" (bar), "head" (latrine), "boon-docks (woods or swamps) and "boondockers." Field equipment was "782 gear," so named for the form that was signed on receipt. Nautical terms included deck, port, starboard, overhead, hatch and countless other expressions and words that replaced the civilian vocabulary. Privates were instructed there were only three ways to do things: the right way, the wrong way and the Marine Corps way. Manchester was wounded in 1945 on Okinawa, and years later he wrote of the Pacific War. Mr. Manchester also noted, "Astonishingly, I adored Parris Island." Author Gilbert P. Bailey concurred. "No one ever likes boot camp yet somehow, grimly, you do like it, in spite of hell." Author Robert Leckie offered, "All the logic seemed to be on our side. The Marine Corps seemed a madness."

Private Perry N. Coley arrived on the depot just before the war began. His platoon was billeted in wooden-floored, two-man tents that were furnished with collapsible cots, under which Springfield '03 rifles were suspended by tie-ties. Shoes were aligned adjacent to locker boxes. Bunks were tightly made, and the stenciled name on the end of the recruit's blanket was to be visible from the platoon street.

A single light bulb illuminated the tent. Towels were hung from a rack on the center tent pole, and each recruit's galvanized steel bucket, which was used for washing and cleaning, was placed outside. The tents could be heated with a small stove. In the summer the tent flaps were rolled up to permit the circulation of air. A platoon's drill instructors lived in a private and separate tent facing the platoon street.

Coley recalled the initial shock, the strict discipline and the South Carolina heat even in the later months of the year. Drill was taught on the sand parade field. The conspicuous tracks of recruits' "boondockers" (field boots) imprinted a horseshoe design that Perry later recalled on Guadalcanal. One always knew when a Marine passed by the imprint of that sole design.

Campaign hats were issued at this time; however, pith helmets could be worn during the summer months. A rifle cartridge belt, bayonet, canteen and first-aid pack completed the training attire. Drill instructors then dressed in long-sleeved khaki shirts and trousers, while their swagger sticks "seemed to be one symbol of their authority."

After the war, Coley joined the newly created United States Air Force, from which he eventually retired an officer. Major Coley was proud of his Marine service and never really left the Corps in spirit.

Private James C. Guest enlisted in the Marines as Ignatius Conrad Goszczynski in October 1941. His name caused him problem in boot camp from the very start.

The very first time our drill instructor began to call the roll, I was ready. When he ran through the last of the names…I was all ears and as tight as a finely tuned violin. When I saw him look up and down the roster, then look down at it again and seemingly sneeze—I answered "Here Sir!" He looked at me, asked me to pronounce it, which I did, and he put an end to my anticipated, repeated torture when he announced, "From now on you will answer to the name of Murphy."

Goszczynski retained the name "Murphy" throughout boot camp, and legally changed his name to Guest in 1964. But "Murphy" experienced an even greater ordeal than his name crisis, which only he can describe:

On the first or second day, among other orders, instructions, chew-outs and admonitions we were informed that any of us who received three or more pieces of mail in any one day would have to go thru the "belt-line." The "belt-line" consisted of your entire platoon lined up in two rows, an arms' length apart facing inward separated by a space approximately four feet wide. The "scapegoat" then ran the gauntlet from one end of the double line to the other, while those forming the "belt-line" swung their doubled over belts at your buttocks as you ran by. Unfortunately, one of my buddies, instead of holding the two ends of the belt in his hand and striking with the looped end of the belt, did just the opposite!

As I approached the end of the "belt-line" I experienced an agonizing, sharp, sudden pain in my scrotum. The belt buckle had come around my upper thigh and struck my sac. I almost passed out from the pain, the DI was upset and nervous not to mention worried. All because I had that date,

CHAPTER SIX

> *Oct. 29, my birthday anniversary, received two birthday greeting cards and one letter. It had been impossible for me, with the demands on our time while in basic training, to write my friends and relatives to inform them to arrange a schedule for writing to me!*
>
> *The next morning the scrotum, which is normally creased and wrinkled, was so swollen the skin virtually shone while it sported a veritable rainbow of colors, ranging from pale blue, through several shades of green to several shadings of brown. Naturally, I refused to go to the sick bay for fear of being taken out of my platoon, an undreamed of possibility in those dear, departed days of our ignorance and culpability.*

Private "Murphy" remained with his platoon despite the pain and discomfort from the lingering wound that remained with him when his platoon moved to the rifle range. Snapping-in "almost proved my undoing," and brought even more excruciating pain. Rifle qualification, mess duty and the final days of training ensued for platoon 161, which completed its training on December 7, 1941. "Murphy" recalled that, he heard about the Japanese attack on Pearl Harbor from his DI and that the instructor replied, "The G--damned Navy will never live this one down!" Christmas 1941 was not the best of times as rumors filled the base. Some anticipated that Germans spies would land on the Carolina coast. Recruit training mostly ignored the holiday—a war was on.

Private James Sherrill recalled that his first week on the island was one "bit of terror after another." Enlisting in the Marine Corps seemed to be the greatest mistake of his life. The shouting and mental intimidation were the worst, but after a week or so the platoon learned to live with their ordeal. Sherrill, who lived in barracks and in two-man tents in 1942, recalled the sand drill field, sand burrs and the ubiquitous sand fleas. His rifle coach "was a most compassionate man"; however, the drill instructors were quick to punish a platoon by double-timing. Recruits who dropped their rifles were required to sleep with the weapons. Sherrill seldom saw an officer. Sherrill lived for the platoon mail call. He did not resent the demanding DIs, whose "discipline instilled in boot camp is what took most people through the war."

Mr. George McMillan listed a number of publications to his credit. None were more prized to the author and World War II combat correspondent than *The Old Breed*, a popular history of the First Marine Division in World War II. It was my pleasure to have known him. McMillan first heard the expression "you people" from a Yemassee Junction receiving sergeant who looked in disgust at the future Marines. "So this is what God has sent me," was the sergeant's kindest remark as the men boarded the train for Port

Parris Island

Royal. McMillan was barged to Parris Island in 1943, and then herded to the recruit receiving barracks. Following an issue of personal items and 782 gear, the drill instructors marched the new platoon to their metal Quonset huts, amidst chants of "You'll be sorry. You'll be sorry," offered from other Marines. McMillan's life was mostly confined to the drill field, where things were always done with one's rifle carried at high port. DIs stared eyeball to eyeball at recruits, screaming obscenities at them all the while.

McMillan's DI Sergeant Brown, ignored the training schedule and threatened, "I'll drill you till somebody drops." The DI seemed prejudiced against platoon high school "grad-U-ates," as he delighted in drawing out the word. Sergeant Brown also delighted having the platoon crawl in the mud and "scramble like eggs." At other times a drill instructor would strike a recruit on his pith helmet with a swagger stick. McMillan recalled, "We did not feel we were being put upon" any more than any other platoon.

Along with the accelerated drill and rifle instruction, Private McMillan remembered letters and newspapers flying through the air at mail call. Recruits were never permitted to speak to members of another platoon. To do so might result in lapping the parade field, strenuous exercises with rifles or "riding" (scouring) a mess hall range. A recruit who referred to his rifle as a "gun" could be required to write the following sentence five hundred times: "There are 50,000 Marines who have rifles, but I'm the unlucky son-of-a-bitch who has a gun."

Quarters urinals had to be spotless, and utility training uniforms were washed so frequently they seemed to turn white. Many recruits suffered from heat rash, which was treated with calamine lotion. Favorite shower songs set to popular melodies were "Tell Your Troubles to the Chaplain" and verses of "When the war is over we will all enlist again, *in a pig's asshole we will!*" McMillan's greatest fear was that he might be set back in the training program or fail to qualify on the rifle range.

It must be remembered that the number of training weeks were reduced during World War II. Island visitors were few, depot security was strict and recruit base liberty was rare. Air conditioning was unknown and most contact with the outside world was only received in a recruit's mail.

Private John C. Stevens II and a group of about fifteen others rode the train from Washington, D.C., to Yemassee, South Carolina. There was little to eat, and a shortage of coach seats required some of the men to sleep in coach baggage racks. Another train took the group from Yemassee to Port Royal, where they were loaded on to large trucks. "From Yemassee onward our civilian days were over, as there we started to get the 'word.'"

At thirty-two years of age, Stevens arrived at Parris Island in December 1943. He recalled the frenetic training pace and the "blasphemous

CHAPTER SIX

encouragement and threats and eating an unforgettably poor meal." Platoon 906 also experienced the ordeal of the hygienic building, clothing issue and the Marine Corps haircut. Stevens had the good fortune to see Master Gunnery Sergeant "Lou" Diamond in the Hygienic Building, "an old timer and hero of the first part of this war." The training day was from 0430 to 2200, and Private Stevens listed the events:

0430	arise, shave, shower	1330	close order drill
0500	chow	1530	chow
0530	sweep, swab, clean up	1630	field day or lecture
0600	setting up exercises, boxing	1830	reading, writing, study time
0700	close order drill	2130	hit the sack
1030	chow	2200	taps
1130	lecture		

Three weeks were spent at the rifle range where Stevens platoon lived in hastily constructed PB (Personnel Barracks) buildings with outside toilets. Little time was left to fret about the cold, since the rifle range training was all consuming of time.

0445	arise, clean barracks	1630	steam cleaning racks
0515	chow	1700	chow
0600	physical drill under arms	1800	drill
0700	range	1900	field day
1100	chow	2000	shave, shower, write letters
1200	range	2200	taps

Snapping in was an ordeal. Private Stevens qualified as a marksman, having dropped from a sharpshooter the previous day. A week of platoon mess duty followed the rifle range, but Stevens was assigned to work at a property shed. Part of his duties included visiting the Parris Island farm to "get a load of cow manure and bringing it back to spread on the grass around the barracks." The completion of his training followed, and in February 1944, Stevens departed Parris Island for ten days of recruit leave. A son, Judge John C. Stevens III, was a 1957 Parris Island recruit and author.

New York State Supreme Court Justice Donald Mark recalled recruit training in 1944. Private Mark was petrified of the two corporals and a PFC who were his DIs. "I thought I was in great physical shape, that I would have it beat, but I never knew how tough mentally it would be."

Parris Island

Recruits were permitted "free" mailing privileges from "Marine Bks Parris Island, S.C.," and Private Bernard William "Bill" Cruse Jr. made the most of the privilege in 1945. In a succession of letters accompanied with graphics, the recruit wanted to hide his homesickness and frustration while trying to convince himself that joining the Marines was the correct course of his life. As Cruse adjusted to the training routine, his letters contained requests to his parents to send him writing paper, reading material and Noxzema lotion for treating sunburn. Other discomforts and illnesses included a sore throat, a virus, heat rash, dental problems, a submucous recession and a bout with Parris Island malaria. On one instance the recruit joked about a huge mosquito. "I had to use my bayonet on a PI mosquito who looked at my dog tag to see what type blood I had."

The drill instructors were demanding and would "drill us in the sand." Snapping in on the rifle range was a "hand-down from medieval tortures," and the only way to escape any training was by going to church. At the rifle range Cruse jerked the trigger of his M1 rifle and was punished by having to stand on a box, hold the weapon on the backside of his hands with extended arms and recite the "My Rifle" creed. Two of the most shocking incidents that Cruse experienced were the accidental killing of one recruit by another on the pistol range and witnessing a suicide by a despondent recruit who blew off the top of his head using a .30-caliber M1 rifle.

Mess duty did little to improve Platoon 332s morale. Cruse was a "bread man" and served on the "firing line" (serving line), becoming aware of his brief authority over other recruits. "Move dust, Joe," was a common expression, and Cruse felt that cursing was a requirement of mess duty for a boot.

Recruits tasted a brief period of freedom when the war ended on V-J Day. The victory over Japan saw numerous celebrations. Between 18,000 and 20,000 recruits and others gathered at a World War II base Outdoor Theater to hear prayers and a speech by the Parris Island commanding general. Planes from Beaufort and Parris Island flew in V formation over the depot several times. Recruit training momentarily paused to mark the end of World War II.

Private Cruse lost interest in becoming an officer candidate toward the end of his training, and anxiously looked forward to his boot leave. He was eventually sent to China, where he saw people executed for stealing a bar of soap. He witnessed other unpunished inhumane crimes whose images remained with him for the remainder of his life.

Fate was also unkind to Cruse. On a truck convoy between Tientsin and Peiping the Marine was "banged up pretty well," receiving a facial injury that required six operations in a Veteran's Hospital. In his fifty-fifth year,

CHAPTER SIX

the former Marine stated that he would be the first to volunteer for service in a national emergency. But "if I had to go through PI again, when they got me back they'd get three damn good men, including the two it takes to drag me back."

Strange things occur in boot camp.

Parris Island

Recruits have at times worn lightly colored helmet liners in an effort to prevent sun strokes and other heat related injuries.

CHAPTER SIX

Parris Island

Drill has always been a significant part of Marine Corps boot camp.

A drive in movie screen was a prominent parade field (deck) structure for many years.

Chapter Seven

ONE OF THE OUTSTANDING CHARACTERISTICS OF THE RECRUIT DEPOTS IS THEIR ABILITY TO EXPAND AND CONTRACT IN SIZE AS TIMES AND CONDITIONS REQUIRE.

PARRIS ISLAND BOOT, 1947

World War II saw the California facility greatly expand. Recruits in the late 1940s recalled that the Navy yard and a hospital were just across the fence from the Marine Corps base. There were few beauty spots, and the drill instructors did not allow those in their charge to be too curious. Some recruits sneaked a look at the cars and airplanes, but a woman was seldom seen.

First Sergeant Charles "Charlie" Carmin (USMC Retired) is a Marine's Marine. I casually knew Charlie as a fellow Parris Island drill instructor, and years later as an e-mail friend. The first sergeant enlightened me during our friendship of his illustrious Marine Corps career.

Carmin was a Kansas farm boy who was awed by the World War II Marines. He first attempted to enlist in the Corps at the age of fourteen. Charlie had problems with a high school teacher at eighteen and wanted to quit school again. Promising that he would finish his education in the Marine Corps, Charlie's parents consented to his ambition to enlist. The decision "changed my life forever."

Charlie hitchhiked to Wichita, Kansas, to inform a recruiter, "I want to be a Marine." Following examinations and taking the oath of enlistment, Carmin and several others were soon enjoying the comforts of a railroad Pullman car, and dining in a fashion that was a new experience for Kansas farm boys. San Diego was reached in two days, where the lad was met by Marines in the railroad station to escort him to the San Diego Marine Corps Recruit Depot. The shock treatment began.

"Drop your cocks and grab your socks" was Carmin's wake-up call. All of the recruits were ordered to make up their bunks and to "get your asses in the head." Charlie followed the others not knowing the meaning of a "head." "Only then did I realize that a head was a civilian bathroom." The recruit had hardly exited from the head to discover that his drill instructor had destroyed his bunk. "You dumb shit! You better learn how to make a bunk the Marine Corps way or I will have a piece of your civilian ass." Stress teaches one how to make a Marine Corps bunk.

Some recruits were confused at the initial drill session, not knowing their left foot from their right. The senior drill instructor informed the new recruits that they were the "sorriest and dumbest bunch of civilian assholes that he ever had the misfortune to see." All civilian clothing was removed and mailed home as the recruits stood naked in a shed with only the cover of a small towel around the waist. Haircuts and clothing issues followed. The military clothing was stuffed into a large canvas sea bag. On the command of "right shoulder, sea bags," the "herd" moved to their barracks, where the drill instructor's tirade never ceased. "All boots are lower than whale shit which is on the bottom of the ocean. You march like a bunch of girls and won't make a pimple on a good Marine's ass." Private Carmin's desire to be a Marine only increased.

The recruit discovered that his barracks were relatively nice for his circumstances with a cooling ocean breeze. However, Private Carmin's Platoon 22 could do nothing right. Charlie had never been to a dentist, and it was discovered that he had five cavities. Carmin stiffened in the dental chair, gripped the arms and saw smoke egress from his mouth. The Navy dentist offered, "That doesn't hurt, Private, 'cause you're a Marine. I didn't holler or pass out but was sure mad at that SOB. I've hated Navy dental people ever since."

Instructions soon followed in all basic military subjects, but the recruits were never addressed as "Marines." Their most commonly used names were "Boot, Shithead, Girls, Pussies and Maggots." The officers addressed the boots as "Recruits or Privates."

All was done with bugle calls at the recruit depots. At chow (dinner, etc.) 500 to 600 recruits entered a mess hall and were ordered to stand at attention at their table of ten. Once all were in position at their respective tables, a drill instructor commanded, "SEATS!" "Everyone had to sit down at the same instant. If all did not move in unison the DI would call us to attention and say, 'I want to hear five hundred asses hit those damn seats at the same time.'" When the drill instructor was satisfied, he ordered "EAT!"

When Charlie joined the Corps and went through the San Diego boot camp they served family style in the mess hall. All of the food was placed on the tables in large bowls and the plates and eating utensils were on the table. There was a lot of clatter when the recruits turned the plates over and

CHAPTER SEVEN

started reaching for the bowls of food. "You had to be in the right place at the table for if you received the bowl last the ration was pretty small. You could throw away some things, but not the meat and vegetables, just the bones and paper."

Cigarette smoking was a privilege. Carmin's DI used a barracks red standing light to inform the platoon when they could smoke in a designated area. When the light was on, the "smoking lamp" was lit. If a recruit was discovered smoking when the red light was off, a pack of lighted cigarettes was placed in the recruit's mouth, a galvanized steel bucket was placed over the culprit's head and a towel was wrapped around the neck that trapped the tobacco smoke. The DI tapped on the bucket with his swagger stick until the instructor safely ended the lesson of obeying orders. On other occasions a recruit could be required to smoke a pack of cigarettes inside a metal wall locker. Another obedience lesson was learned. A recruit found chewing gum or receiving it in the mail was ordered to chew entire packs.

The recruits slowly became Marines as their training continued. That evolution was assisted by athletic competition with other platoons. Disassembling and assembling weapons was another form of competition, as was erecting shelter halves (tents), boxing bouts and push up contests. There was always drill and any failure to be attentive could be corrected with a DI's kick on the shin. It was all planned to instill discipline, obedience and esprit de corps. The recruits gradually learned that they could endure unknown hardships and developed a "can do" pride.

At the end of the third training week the platoon was moved to a rifle range about twelve miles north of San Diego. Camp Matthews had tents with wood decks, and the ground was devoid of grass and trees. "The dirt and sand blew constantly. The only good thing about this area was a cool evening breeze coming in off of the Pacific Ocean."

Rifle inspections were frequent. If a dirty rifle was detected, the recruits were ordered to fall out with their weapons and their full sea bags of clothing, to be marched to a hill or a sand dune. The recruits ascended "Agony Hill" with their rifles slung over one shoulder, and their heavy and awkward sea bags hanging over the other shoulder. The platoon was ordered, "To the rear, MARCH!" as the hilltop was reached. The process was continued until Platoon 22 fell exhausted and scattered on the ground.

Private Carmin and the others were taught "the significance of teamwork. What your buddies or you failed to do could hurt the entire platoon. It was a principle constantly stressed throughout my Marine Corps training. The unit and the mission always came first."

Platoon 22 anxiously approached their significant rifle qualification day. One drill instructor threatened, "If any of you assholes fail to qualify tomorrow as shooters, I'll make you shit in your buckets and throw it at one another." The recruits believed their drill instructor, and Charlie qualified. Two nonqualifiers had to wear their clothing backward and were marched separately from the platoon.

More training ensued when the platoon returned to their home battalion. By now the recruits were "beginning to get the hang of it," and the training did not seem so harsh. The drill instructors might be human beings after all. The recruits learned that their senior DI was captured on Wake Island early in World War II. Four years were spent in Japanese prison camps, and a facial scar appeared to be a knife or a sword wound. Sergeant Gardner also had a tattoo of "a large snake's head coiled around his torso with the last coil wrapping around his neck. The snake's head with its open mouth centered the chest." The rumor was that the drill instructor received the tattoo while a prisoner of war.

One recruit "went over the hill AWOL," (absent without leave) and was later apprehended. The recruit was sentenced to hard labor. In the late 1940s and early '50s, court-martial sentencing was performed in front of formations of troops. The convicted person had a large white P painted on the front and rear of a green utility jacket that was devoid of any rank. The charges and sentencing were read aloud by the Officer of the Day. At the conclusion, the troop formations were ordered, "About FACE," as "all hands turned their backs on the prisoner as he was marched away." Private Carmin found the lesson impressive and, "I certainly didn't want to go to the brig."

One other recruit was a thief. The boot was discovered and given the opportunity of a court-martial or running a platoon belt line. The boot opted for the belt line and walked through a corridor of sixty-five recruits striking the culprit's buttocks and legs. The humiliation lingered but the shamed thief completed his recruit training minus any record of his disgraceful deed.

Additional field training and more uniforms were issued during the final training weeks. One uniform was the coveted dress blues. Graduation was "one of the proudest days" in Charlie's life.

The numbers of recruits arriving at Parris Island greatly fluctuated after the war until 1950, when the Parris Island and San Diego recruit depots experienced a rapid expansion once more. Moreover, Parris Island welcomed the permanent return of the female Marines upon President Truman's signing of the Women's Armed Services Integration Act. Parris Island was designated as the single female recruit training installation with Platoon 1 graduating in March 1949. As in the past, the female Marines freed males for combat duties in the event of a future war, which in 1949 was not far away.

CHAPTER SEVEN

The women's training period lasted for about six weeks with approximately forty-eight recruits in the platoon. The training consisted mostly of mastering clerical duties, but the ladies indulged in hours of drill and general Marine Corps education. They participated in such male exercises as touring the tear gas chamber while singing "The Marines' Hymn".

Black recruits did not train at Parris Island during World War II. They had segregated camps at Montford Point, Camp Lejeune, North Carolina, where they were initially trained by white drill instructors and officers until black DIs were assigned the task. The Montford Point Program was the same as the normal course for whites, but racial barriers did exist. Moreover, black DIs were as demanding as the white instructors, and one black recruit recalled that discipline seemed to be his instructor's lone stock in trade.

The Montford Point camp was closed in September 1949, when all black recruits began recruit training at Parris Island and perhaps San Diego. Although they were initially assigned to segregated platoons their small numbers made this arrangement unsatisfactory. Economic and new military racial policies demanded total integration. Black women were accepted into the Marine Corps in 1949 and trained exclusively at Parris Island.

Several base and temporary drill instructor schools were formed in the late 1940s, as drill instructor training was more refined from 1945 to 1949. Lieutenant Colonel Robert D. Heinl Jr. argued, "The Marine Corps will never be any better than its recruit depot DIs." To Heinl drill instructors were the "keepers of tradition," and the "architects of the Corps" who should be no less than "the very cream and elite of all our noncommissioned officers, selected on a Corps-wide basis."

Male recruit instruction was much the same as it was during the war. The Yemassee Receiving Station remained "The Gateway to Boot Camp," the last civilian stop for recruits en route to becoming Marines. A ten-week program was altered in 1948 to include longer hours of M1 rifle instruction, additional drill and guard duty, organized athletics, physical training, more inspections and classroom instruction by Quantico-schooled special subjects instructors who assisted the DIs.

Sergeant Arthur Kozak, who began World War II on Midway Island, was a Parris Island DI in 1947. Kozak recalled a DI school, but said it was really nothing more than a pool for drill instructors awaiting assignments. The sergeant was fortunate to have an experienced junior drill instructor. In one instance, after a third training week, Kozak took a platoon through boot camp alone because of the acute shortage of DIs.

Drill instructors often dropped and joined platoons in training, as the program remained one of lightning-like speed. The obstacle course was run, the gas chamber was visited, and field entrenchments and mortar positions were viewed. Recruits scrubbing mess hall garbage cans were dubbed "The Man in the Iron Mask." The only recruit relief was church call, the mercy of a drill instructor and pith helmets to shelter recruits from the South Carolina sun.

Recruit dress resembled that of prisoners when the green dungaree jacket was not tucked in the belt and the trousers were not bloused. The World War II "boondocker" shoe was constructed with the rough side of the leather to the outside of the ankle-high boot. Private Julius J. Ginther recalled in 1947, "In order to break them in, the DI had us put the shoes on and go to the head and put each foot in the toilet, flush it and get the boondocker soaked. Then we were taken out on the parade ground and troop and stomped until the shoes dried. It seemed to work."

The three weeks that were spent on the rifle range remained a high point for most recruits. A new M1 rifle course was introduced in 1948 having a maximum score of 250 points.

The practice of promoting select recruits to private first class was resumed in 1947, as was the ten-day recruit leave. But before these treasures were won, recruits had to earn the right to be addressed as a "Marine." That price included barbers who denuded recruits' heads, clothing issues being hurled at recruits and drill instructors who seemed to enjoy running boots in and out of squad bays at breakneck speeds. Inoculations, sore feet, barracks field days and inspections remained part of a recruit's life. Mail call was an athletic event when recruits had to sprint for their treasured letters as they were sailed across squad bays and over the parade field. It was not uncommon to see a platoon of recruits wearing buckets on their heads, because some "knucklehead" called his rifle a "gun."

Private Thomas H. Patten Jr. never forgot the 1946 sack drills, push-ups, cursing and recruits holding rifles over their heads. One mass punishment occurred when a drill instructor closed all of his platoon's squad bay doors and windows and turned the steam radiators up to full heat. Years later Patten applauded the training that was, in his opinion, necessary to make tough Marines.

Private Don Betz arrived on the depot in 1948. He was greeted by several unflattering remarks from Marines he thought suffered from combat fatigue, since "no one would talk to people like that." A first cigarette was allowed on the second training day, and personal possessions such as dice, cards, knives and lewd photographs were confiscated from the new enlistees. Any weapon or alcoholic product, including after-shaving lotion, was taken from recruits.

Betz was quartered in a metal Quonset hut with twelve other recruits. All of the boots thought it strange that drill instructors carried no weapon

CHAPTER SEVEN

in their leather pistol holster that was attached to a thick cloth duty belt. The web belt and holster was at this time the authority symbol of the DI. Cigarettes and lesson plans were frequently concealed in the holster. Drill instructors stuffed Parris Island toilet paper into the holster bottom to keep it from bending and destroying the mirror shine.

Strict discipline was always demanded, with severe punishment for those boots refusing to do things the Marine Corps way. On several instances Betz, a very large man, was booted "hard enough to lift my feet off the ground" and was struck in the chest. His drill instructor was willing to take on any recruit in a fight, but "no one volunteered."

Private Betz remembered, "We never knew what was in store for us for the day." Yet he and his platoon did not perceive the DIs as cruel. The platoon was proud to carry their red flag with its emblem that designated them an honor platoon. Betz retired a Reserve Sergeant Major (USMCR) who maintained that, "If I could do it all over again, I would not hesitate, and I would not ask that anything be changed."

It must be remembered that drill instructors must work as a team. Disagreements can occur that are usually solved by the senior instructor in command of the platoon, although a senior drill instructor could be at fault or incorrect. This was especially true prior to the officer supervision that is so evident at the recruit depots today.

Tom McKenney enlisted in the Marine Corps in January 1949. Private McKenney was a tall and intelligent young man with a great desire to be a Marine, and a semester of college ROTC behind him. The private respected his two junior drill instructors, who were PFCs, but soon discovered that he could not respect his senior DI, a corporal. Moreover, there appeared to be some deep friction between the corporal senior drill instructor and a more respected PFC junior assistant. This junior DI had a relaxed leadership style, but commanded the respect of the recruits, and they "snapped when he drilled them." In a rather bizarre occurrence, one day the junior DI was no longer there; there was no explanation given—he was simply no longer there, and McKenney's platoon wondered about their junior DI.

> *One day the good junior DI wasn't there; nothing was said to us, he was just gone. Later that day we were in formation out on the company street, and our missing Pfc drill instructor went by, hanging onto the back of a garbage truck. He was waving and grinning at us, with a big, black P (for prisoner) prominently displayed on the back of his dungaree jacket. We never saw him again, and assumed that our "psych senior DI" had "run-up" the Pfc, who was sent to the brig.*

Parris Island

Although Private McKenney held little respect for his corporal, the recruit was appointed as the platoon's guide. On one instance McKenney attended church call, and was that afternoon ordered to paint a barracks stairwell. "This was undoubtedly punishment for my going to the chapel that morning." Moreover, the corporal DI had little respect for chaplains whom he hated.

During his barracks chore Private McKenney overheard a conversation that included a recruit and a DI whom he admired from another platoon.

> *As I painted, the senior DI next door took a recruit aside. The kid's parents had come to visit him on Sunday afternoon, which the DIs actively discouraged, but legally couldn't forbid. The DI was explaining to the recruit, in a fatherly way, why it wasn't good to be visited by parents during boot camp. He explained that he only had 3 months to turn civilians into Marines, that the intense training schedule was time-tested, and that one afternoon with his parents could undo what he had been working to build into the recruit's attitudes and values for weeks. He finished by saying, "I am responsible for what you become, and I am as jealous of you as a mother is with a new-born baby."*
>
> *I sat there, listening, and thought, "Dear God—what I wouldn't give to have a senior DI like that man." He didn't "thump" the kid (like many other DIs would have done), didn't even berate him; he didn't need to. I sometimes wonder if that other recruit is still alive after nearly sixty years, and if he remembers that Sunday night's private lesson from his senior DI.*

Private McKenney had on one occasion considered placing his senior drill instructor on report, for his sometimes-insane behavior, and his occasional cruelty, but decided not to do so. "I think it would have constituted a failure on my part." The corporal was later court-martialed for taking money from recruits, and the drill instructor was obviously unfit to command recruits. The DI corporal took leave when McKenney's platoon graduated, and departed Parris Island on the same bus with some of McKenney's platoon. The corporal was soon drunk and told his just-graduated recruits to call him "Kenny." This further revolted many of his former recruits. Private McKenney experienced a successful career and was commissioned in 1953 at the very end of the Korean War. He was retired as a lieutenant colonel in 1971 for disability incurred in Vietnam.

Former boot Gilbert Bailey revisited Parris Island in 1949, on assignment for the *New York Times* to observe the "skinheads" in "the land that God forgot." The author discovered that boot camp remained "a stern exacting

CHAPTER SEVEN

parent" that put young men and women through an experience that they could "be proud of, something they won't forget."

Recruits still joined the Corps for "foreign legion" reasons, but they quickly experienced "the usual cold-water disillusionments of military service." Drill instructors were demanding as ever, and if a DI told a recruit "to go out and get him a slice of the moon," recruits immediately proceeded outside and "started jumping in that direction." Bailey found that boots had little time for morale problems in the "human production belt," and in the "ordeal type of boot training" that the Corps retained. He also observed that the only speeds on Parris Island were "marching and running," as during the war. The reporter and author additionally witnessed the persistent belief that the enemy was at the Parris Island main gate, and more Marines were necessary to repel them.

Mr. Bailey recorded there were only two types of recruits—those who could shoot and those who could not. Rifle marksmanship was stressed to the degree that "the fate of the nation hung on a bull's-eye." Once a recruit completed his training, the new Marine believed that he was "one of Jehovah's chosen commandos," who had proven himself as one of the best disciplined, best trained, best dressed and "roughest characters who ever donned a cartridge belt."

The *Marine Corps Gazette* reported in 1948 that West Point and Annapolis "may turn out more brains per diploma, but not more discipline" than Marine Corps recruits. The magazine stated the recruit "is drilled, lectured, and guided" for ten weeks, and that he is always told what to do until he is graduated and is all alone. The *Gazette* asked:

> *Have you ever seen a young colt, broken away from his pasture, standing bewildered in the middle of a highway? Blowing your horn will not do much good; you've got to lead him. And so your recruit standing there has been trained to be led.*

And led he was to Korea, where the Marines fought another, but very different and bloody Asian war.

The famous Campaign Cover (hat) was returned to the Marine Corps in 1956.

The odds are not in favor of the recruit in boot camp.

CHAPTER SEVEN

Some recruits have thought that joining the Marine Corps was the worst decision in their life. Years later they are proud to be and have been a Marine. Semper Fidelis.

"Can't" is not a word to be tolerated from recruits in boot camp.

Chapter Eight

> THE THREE MOST IMPORTANT PLACES IN THE MARINE CORPS ARE THE TWO RECRUIT DEPOTS AND THE BASIC SCHOOLS IN QUANTICO...WHERE WE TRAIN MARINES.
>
> GENERAL EDWIN A. POLLOCK

On Sunday June 25, 1950, a three-year war erupted in Korea that took the lives of more than 33,000 of the nation's sons. *Time* magazine reported the Leathernecks had their packs filled with the glory earned in World War II, and were prepared "for anything that can be thrown at them." Strapping on their leggings and packs, Marines fought at the Naktong River, the Pusan Perimeter, Inchon, the Chosin Reservoir and in fierce hilltop battles that included Carson, Reno, Vegas, Bunker Hill and The Hook.

The Marine reputation of being "the first to fight" attracted thousands of recruits to San Diego and Parris Island who were motivated to serve the nation during the Korean War. As in the past, the depots significantly expanded. Tents were erected over much of Parris Island and vacant Quonset huts were reopened to house a tidal wave of fresh recruits and Marine Corps Reserves (USMCR). Selective Service (the draft) was also used.

The training duration was varied from twelve to ten and eight weeks as eight Parris Island recruit battalions were activated. As of March 24, 1952, there were 21,540 recruits on the depot, including 129 women. Other sources place the total numbers at 24,424. Both counts surpass the World War II record of approximately 20,000 recruits who lived on the island *at one time*.

The Marine Corps enlisted women between the ages of eighteen and thirty in a program that lasted six weeks. Although major differences existed in the male and female curricula, the women's course was in its way as demanding as the men's. Yet Major General Robert B. Luckey and Lieutenant General James P. Berkeley offered that any hardening of the

CHAPTER EIGHT

women was undesirable. The "way of life at Parris Island for a male Marine shouldn't necessarily apply to a woman Marine."

A typical schedule for women in the 1950s began with reveille at 0515. Cleaning chores and breakfast followed and a personnel inspection was held. The women attended classes until the noon meal was served. Schooling was resumed in the afternoon, and there was always drill. More study followed in the evenings along with barracks and personal cleaning chores. There was schooling and free time until taps at 9:45 p.m.

Mary-Ann Perry (Fitzpatrick) enlisted in Boston in 1959, and was immediately placed in charge of three other women for their journey to Parris Island by train. The women were given a list of recommended articles, including seasonal civilian attire for their Southern trip. They were also advised to take a housecoat, slippers and a sufficient supply of under clothing for the ten-week period. Other items included toilet articles, a conventional style bathing suit and cap, white socks, a full-length slip and a girdle.

It was recommended that some of these items be purchased during their first Post Exchange (PX) call, and that recruits should have fifteen dollars in expense money to tide them over until their first recruit pay call. Forbidden articles included patent medicines, aspirin, ointments, laxatives and photographs larger than billfold size. Chewing gum or any perishable food was forbidden, as were clocks and jewelry, except for watches and simple rings. Cameras and small radios were allowed, but all were subject to the instructor's rules.

The four women arrived at Yemassee to board a Greyhound bus to transport them to Parris Island, where they were deposited with some newly enlisted men. It took several days for Perry's platoon to form, during which time the women were herded about by male DIs. The male instructors were always present, while female instructors acted as platoon sergeants accompanying the recruits to areas and examinations where all men were banned. Clothing issues and processing followed, but rifles were not then issued to female recruits. Much of their instruction concerned clerical and administration training, and drill was constant, even in the rain. The training also included a visit to a tear gas chamber.

Private Perry discovered that Parris Island was like being in a prison, and that "my life was not my own anymore." Profanity and hazing were common with some DIs correcting the women using their swagger sticks. Push-ups were ordered for conditioning and punishment. Field days made the barracks spotlessly clean. All mailed food had to immediately be dispatched, and smoking was restricted. Talking was not permitted during activities such as those in laundry rooms.

Parris Island

The women were required to wear girdles while in uniforms, but perfumes and cosmetics were not allowed. Locker box keys were worn tightly around the recruits' necks, and regulation underwear was worn, as were cotton hose. Nylons and cosmetics were permitted only after graduation. The bothersome insects did not discriminate against the male or female recruit, and loneliness and isolation caused additional problems. Nevertheless, thirty of the original sixty-eight Platoon 16-A recruits completed their three-month course and won the right to be Marines.

It has been common as previously noted, for San Diego and Parris Island Marines to argue how severe their training was. For example, West Coast recruits are often called "Hollywood Marines." While the California depot must contend with the distractions of a city and an airport, Parris Island has the infamous sand flea. The annoying insects are known as sand fleas, sand flies, sand gnats, biting midges and punkies. Others have humorously called them "ctenocephalis sandis," "no-see-ums" and "flying teeth." They are not fleas, yet some dictionaries list them as such. The correct genus classification is *Culicoides*, and the pesky gnat will "bite a Rebel or a Yankee." Their breeding grounds are in salt marshes and in sand. A female can produce two generations of "Flying Teeth" in one year.

Thousands of boots and other Marines shared the sentiments of Private Bernard William Cruse in July 1945:

> *And then there's the sand-flies, or sand-fleas, whichever you prefer. They don't bite, they don't sting, they don't, as far as I know, carry any disease, but they can sure worry the fire out of a guy. Especially while we're marching do the sand-fleas love to bother us. While we're sweating to keep up, our rifle on our right shoulder and our left hand swinging wildly for balance, we're very disgusted at the world-in-general. And then to have those #=$/* sand fleas; that's almost too much. But once the DI put it this way in his tender manner.*
>
> *"Get your Godamn hand down, scrounge!" "But, sir, the sand-fleas—" "Godamn it, I said keep your hand down! Don't you eat to live?" "Yes, sir." "Well, the sonofabitchin sandfleas have to live too, don't they?" "Yes, sir." "Those damn fleas were put here for a Godamn purpose. They're supposed to bother recruits. That's why they're here! And I won't have any of you knuckleheads smashing the Godamn life out of them!"*

Boots during the early 1950s vividly recall departing Parris Island with the assurance they could endure any confronted danger or task. Boot camp was the standard by which to compare any future ordeal. Three recruits who held these beliefs were Privates Gene Duncan, Gene Alvarez and Jim T. Darnell.

CHAPTER EIGHT

Private Duncan enlisted in February 1950. On the train from Yemassee to Port Royal, Duncan observed a conductor who exaggerated that for fifty-five years he had witnessed recruits en route to the depot. The old-timer enjoyed teasing the enlistees, as he walked through the car singing, "I Wonder Who's Kissing Her Now." At Parris Island Duncan recalled that his first Marine Corps meal was Spanish franks.

The platoon processing ensued, which included administrative details. One corporal told the recruits that he wanted to know if their religion was Catholic, Protestant, or Jewish, adding that, "I don't want to hear any other word but one of those three coming from your mouths." All went well until one new private replied Baptist and Duncan then first witnessed maltreatment in the Marine Corps.

The future officer recalled how he disliked the island's marsh odors, and that he was perplexed by his platoon visiting an outdoor theater movie in freezing weather, because the drill instructor thought that "it would be good for our morale."

Recruits were informed at the rifle range that those who failed to qualify would not be permitted to carry rifles while they were Marines, but would instead be "armed with spears." Private Duncan also remembered that when he was promoted to Private First Class, the battalion commander stated, "You should be as proud of that stripe as I would be of a brigadier general's star." Major Duncan retired from the Marine Corps with two Purple Hearts awarded in Vietnam.

This author recalled in 1950 the isolation and homesickness and realized the importance of mail. The World War II barracks and squad bays were dimly lighted, depressing and worn, resembling what one expected in a World War II German concentration camp. All idle moments were devoted to studying the *Guidebook For Marines*, and learning vital information such as one's Marine Corps serial number, which few Marines ever forgot.

A Marine's general orders were greatly stressed. Sentries and fire watches had to memorize the orders by the first training day, since drill instructors and the Officer of the Day could examine a recruit's mastery of the orders at any time while a boot was on a sentry post. I was the first recruit assigned to my platoon's fire watch duty because my name began with the letter A.

The barracks platoon fire watch also had the awesome duty of awakening the DIs at 0500. This was tantamount to approaching a Bengal tiger at rest. Yet I never recalled a DI striking a recruit. Crowded showers, sand fleas, and constant drill were part of his August to November 1950 training, in addition to other never-to-be-forgotten ordeals that were fully appreciated later in the Korean War and throughout life.

Parris Island

The rifle range remained the most relaxed part of the training, although snapping in while seated in sand burr patches and wearing heavy shooting jackets in the heat had to be endured before qualification day. At the end of that significant experience, one's score was chalked on the back of the recruit's shooting jacket, and nonqualifiers had to march at the rear of the platoon with their rifles slung around their necks. Today the humiliation would be regarded cruel. Mess duty was one of several other boot camp low points.

Platoon 99 looked forward to graduation day. All were happy to remove the rope tie-ties from our necks that safely secured our locker box keys. A main PX call was allowed, a platoon photograph was made, and in November 1950 graduation ceremonies were held in front of the battalion headquarters buildings. Shooting medals and PFC promotions were received from the battalion commander. There was no Warrior's Breakfast, emblem ceremony or a "defining moment" for completing boot camp. The defining Parris Island moment in 1950 was to depart the depot, hoping to never see Parris Island again.

Private Jim "Bulldog" T. Darnell enlisted in 1953, and quickly wondered, "What the hell I have gotten myself into?" His DIs shouted obscenities at the platoon, and uniforms were thrown at the boots during their initial clothing issue. Hours of drill followed on the first training day. One recruit squad leader was as violent as a DI. On one instance, Bulldog came off guard duty after enduring the coldest night he ever recalled. During a morning inspection Darnell's frozen fingers and hand could not negotiate opening his M1 rifle bolt. His DI opened the bolt and had Darnell insert his thumb into the rifle receiver and let the powerful bolt drive home. The 9.5-pound rifle hung from Darnell's thumb during the entire inspection, but his hands remained so cold that Bulldog hardly noticed the pain.

Field days were frequent. At these times all items were removed from the Quonset huts, whose concrete floors were doused with buckets of water and sand to be scrubbed white by the hut's recruits. Before the day's end, every item had to be clean and aligned. Prior to taps the platoon recited the "My Rifle" creed until it seemed their lungs would explode.

Bed linen was changed weekly and clothing was scrubbed with brushes on concrete wash racks and hung out to dry with small rope tie-ties. A recruit clothesline watch patrolled the hanging clothes to assure security. Darnell observed that the strict regimentation caused several of his platoon's recruits to be "washed out," but most of them made it to graduation day never forgetting their days as boots.

Private R.R. Warner remembered his galvanized bucket during his 1955 recruit training. The bucket was recalled nearly thirty years later in

CHAPTER EIGHT

Leatherneck magazine as the second most important issue in boot camp that was outranked only by the rifle.

It has been a long time, and memories fade, but I remember that old bucket. It was the first thing issued to me in boot camp, even before I received my gray cotton sweatshirt and utility cap. It came to me filled with towels, a toothbrush, a double-edged razor with blades, soap, a tube of shaving cream, a scrub brush, clothesline ties, and other sundry items necessary for cleaning and grooming. The bucket wasn't new, but it had a well-cared-for and slightly forbidding appearance.

At first I didn't pay much attention to my bucket. I had more pressing matters on my mind – like figuring out how to become invisible whenever my drill instructors were in the vicinity. (As I remember, they were always in the vicinity.)

My inattention to such an important object as my bucket was corrected during the next 13 weeks, and my bucket and I became great friends.

The first thing we did together was stumble toward our new living quarters where the DI, with great care and precision, instructed me on exactly where my bucket was to be placed under my bunk. Then, being certain I had absorbed this, he further showed me how to fill my bucket with soap and water, and how to empty it properly while swabbing the deck or my Quonset hut from one end to the other.

My bucket and I shared other activities, too. When I needed a comfortable place to sit, the first thing I thought of was my bunk. However, my bucket reminded me that was a b-a-d thing to do, and it would practically turn itself over so I would have a good place to sit while shining my boots or cleaning my rifle.

My bucket was many things to me. It replaced my mother on washday. My bucket and I would march together to the laundry rack where, with the help of my trusty scrub brush, we would proceed to remove unauthorized dirt and sand from my military garments. Then, my bucket filled with very clean clothing, we would proceed to the clothes drying area where I would tie my wash onto the lines.

My bucket also replaced my former neighborhood garbage man. Many times we would fall out together to police the area of all loose scraps of non-government property. We seldom found anything, but my DIs insisted that those scraps were out there, and we employed diligence in our search. My bucket also replaced our family gardener (dear old Dad) when we got together to weed and care for the plants in the battalion area.

One of the most exciting times my bucket and I shared was when it replaced my high school gym teacher. It was fun seeing how long I could hold it at arm's length. One of our favorite games was running at the double, while I made sure the sand in the bucket never spilled. My bucket needed that sand. I know, because my drill instructor told me so.

But you mustn't think my bucket and I shared only the good times. We had our serious moments, too. It was always ready to serve as a classroom seat when I needed to study my Guidebook for Marines. And it was there when I wrote home explaining how much I was enjoying my self at summer camp.

My bucket even helped advance my love life, providing me an amorous perch when I wrote my sweetheart, ensuring her that her "tiger" would soon be home, and she could brag that her boyfriend was a U.S. Marine.

I learned to love my bucket. Together we hauled hundreds of thousands of gallons of water (or so it seemed at the time). We helped redistribute all the sand at boot camp several times, and we made sure no trash, cigarette butts or chewing gum wrappers ever stayed on the deck more than five seconds. We cleaned the area from stem to stern. We built muscles together.

In exchange, the only thing my bucket asked for was to be kept clean and shiny, to be kept filled, to be kept empty, and to have the chance to serve its country. This it did with great distinction.

During my tour with the Corps, I was introduced to many other buckets, and we did good service together. But they were strangers, destined to pass through my life quickly and be forgotten. They will never hold a special place in my memory like my old boot camp bucket.

I don't know where my bucket is now, but I'm sure it gave many years of honorable service to several generations of new recruits. I hope it was able to enjoy a well-earned retirement, undented and unscratched.

To my old bucket—here's suds in your eye...

Former Marine and writer Don Mason recalled his boot camp experience in this following interview:

In the summer of 1953 I turned seventeen and joined the United States Marine Corps Reserve in Oklahoma City. I attended meetings at the 8th Rifle Company, and in the summer of 1954 the company was ordered to Camp Pendleton, California for one month of combat training. That experience convinced me that I wanted to go on active duty as soon as I graduated from high school.

I recall being given the choice of which Marine Corps Recruit depot I wanted to train in. I chose Parris Island because I had heard more about it than I had San Diego, and I assumed that Parris Island was the toughest place for recruit training. I was not disappointed.

CHAPTER EIGHT

On February 15, 1955, I departed for Yemassee, South Carolina, a staging area for wannabe Marines heading for Parris Island boot camp and only a short bus ride away. For thousands of men, Yemassee was their first taste of the Marine Corps, and being introduced to tough drill instructors.

At Parris Island a DI yelled for us to get off of the bus and to "fall in" on a white line. At that time the yellow footprints known today had yet to be used. It was late at night and we all had been traveling for some time. But the drill instructors who "welcomed us to their Marine Corps and their island," cared little about how we felt. Much yelling and harassing went on for about thirty minutes before we were ordered to follow the DIs in to some sort of "herd formation" to our Quonset huts. Three weeks later we occupied World War II wooden barracks.

Our hair was soon removed to the scalp and uniforms were issued. We also received a bucket issue that contained a scrub brush and laundry soap. In the 1950s we washed our clothes at an outdoor concrete wash rack and hung the clothes to dry with small pieces of cord called tie-ties. A clothesline watch guarded the clothes. Recruit also pulled fire watch duties walking the barracks for a two-hour tour. As our training progressed we visited dental and medical clinics, and a "head doctor" who interviewed recruits.

I never at any time witnessed anyone in our platoon receiving any severe hazing or beatings. Now and then a DI would smack you with his swagger stick, but never hard enough to bring blood. A slap or shoving you enough to knock you back a few steps was about it in our platoon. I did witness some beatings taking place in other platoons. Our Senior DI did not allow any of his Junior DIs to strike us severely. He served in WWII from Guadalcanal through the war and later fought in Korea. He knew how to get what he wanted, and in the end we were a Depot Honor Platoon.

Our punishment was mostly push ups, running laps around the parade deck, holding the heavy M1 rifle straight out until your thought your arms would break, etc. Bunk drill was also known, where a recruit does drill when lying in their bunk. An exercise known as "toes and elbows" was my misfortune on several visits to the DI room.

A recruit could never enter the DI room or speak to a drill instructor until loudly knocking on the wood hatch [door] frame and clearly reciting, "Sir! Private Mason requests permission to speak to the drill instructor, Sir!" If allowed, the DI responded, "Speak!" DIs were never addressed personally, but were always addressed as "Sir." Recruits were called many names.

Our training also included a gas chamber visit, and we were ordered at the last minute to remove our masks to be exposed to the gas. The

PARRIS ISLAND

lesson was learned to always wear a mask in the event of a gas attack. We also bivouacked overnight, and the sand fleas were terrible. One was never allowed to kill a sand flea. Discipline and taking orders without hesitation were stressed throughout our training. The mental training was very tough.

Many hours were spent practicing close order drill. The Marine drill instructors had a different cadence than any other service, and I never forgot the call. Years later I wrote a poem that was published in Leatherneck magazine.

The DI's Call

I can hear his voice from across the years
And it beckons me back again
To march on his field of long ago
That deck from way back then.

His haunting song never leaves my head
His voice demanding my all
The cadence and rhythm echoes on
The cadence of the DI's call.

Once you've heard his words and song
And the sound of pounding heels
You will always hear it coming back
Across those distant fields.

March sharply you Marines that went before
Don't bobble or bounce or fall
What you hear from long ago
Are the sounds of the DI's call.

Our platoon moved to the rifle range for three weeks, but had only one day to qualify. There were no second chances. Those who did not qualify were harassed and yelled at and in some cases had to wear their uniform backwards for a few days. In our day every one had to fire the rifle from the left side of the piece. If you were left-handed that could be a problem, but that's the way it was. Some lefties qualified and some did not. I qualified with the rifle and that was an honor for me. I felt sorry for the few nonqualifiers, but perhaps they later did better. In Marine boot camp the recruit is under much pressure.

CHAPTER EIGHT

Mail call was the high point of our day, being our only connection to the outside world. We could not have a radio or TV and we had no idea what was happening beyond the Parris Island gate. The only women we saw were female Marines, and then from a distance. I can't recall seeing any civilians. As a rule we had some free barracks time on Sundays, and that's when we answered our mail. But, if we didn't get to it, there were times we would write letters after taps under a blanket while using a flashlight. If caught there was hell to pay.

I learned a lot from the Marine Corps, and especially from my boot camp experience. I made the transition in boot camp from a wet-behind-the-ears young man to a full-blown adult ready to meet challenges as they came along. I am proud to have served in the 1950s. We went before and today are the touchstone to the past for the younger Marines. Our pride and honor remains true to God and country. Semper Fidelis.

Other recruits recalled the commands, "Count Cadence, COUNT, Delayed Cadence, COUNT," and the perplexing, "To The Winds, MARCH!" Also, "STRUT, STRUT, STRUT, HEELS, HEELS, HEELS," which was criticized by U.S. Navy doctors concerning injured feet. But the strutting produced marvelous drills and was good business for the base cobbler shop. "Joe Blow" was a name used for general purposes as was the once popular "Johnson Bar" that was applied to inanimate and other objects. Before recruit smoking was disallowed, a lighting of the "Smoking Lamp" was used by drill instructors to punish or instill morale. Chronic smokers would at times arise in the middle of the night to sneak a smoke in a nearby Dempsey Dumpster, that was a large metal receptacle for dumping trash. Some DIs became aware of the nightly maneuver and met a chronic smoker inside of the garbage container. It was a religious experience for the recruit.

Rifles had to be immaculate for a platoon's final field inspection. Washing the parts in soapy water was permitted, to be followed with the application of a very light coat of oil. Some drill instructors used gasoline for sanitizing the rifles, which was more efficient than the soap and water. The use of flammable materials in a barracks was disallowed for obvious reasons. But with platoon sentries on the alert for the patrolling Officer of the Day, and extreme caution, a platoon could post a near perfect final field rifle inspection by using gasoline.

Some drill instructors encouraged recruits to contribute gifts and money to their DIs, which was known as "Flight Pay." Giving money to superiors, especially if encouraged, resulted in courts-martial that wrecked drill instructor careers.

Parris Island

Recruit and drill instructor vocabulary is extensive, and changes with the times. For example, "gung ho" was once popular, rather than "ooorah." The latter expression makes little sense to older Marines. Conversely, "gung ho" in Chinese means to work together. The expression has become a part of the general vocabulary that is still heard today.

Recruits have lived in tents, barracks and huts.

CHAPTER EIGHT

A tent city faces roughly hewn single-story wooden barracks.

The side tent flaps could be raised to allow cooler air into the tent. Tents such as these were widely used during the Korean War.

The old and new style barracks. The smaller, white wooden building was the last of the First Battalion H and World War II–era barracks. Notice the much larger modern brick barracks.

A modern Parris Island brick barracks.

CHAPTER EIGHT

"Old Corps" recruits at wooden wash tables. Rifles and other equipment could also be cleaned on the tables.

Recruits use later concrete wash racks. Notice the recruits are wearing swimming trunks, and are adjacent to one of the World War II wooden barracks.

Barracks PT (physical training). Recruit training time is valuable.

Taps are eagerly awaited by recruits in training.

Chapter Nine

RIBBON CREEK HIT THE MARINE CORPS HARDER THAN ALMOST
ANY OF ITS FAMED BATTLES IN WORLD WAR II.

NEW YORK TIMES, 1958

Recruit marsh and tidal stream marches most likely occurred since 1915. Colonel Mitchell Paige's platoon was allegedly marched into a tidal creek in 1936. Robert Alan Arthur reported that "Claude the Brutal" took his recruits into the water while wearing buckets over their heads three times in 1942. Robert Leckie observed a World War II DI march a platoon toward the water. If any recruit faltered, the DI bellowed: "Who do you think you are? You're nothing but a bunch of damned boots! Who told you to halt? I give the orders here and nobody halts until I tell them to." At the crucial moment the drill instructor screamed, "Come back here, you mothers' mistakes!"

Privates Don Hetzner and Jim Darnell's platoons were marched "through a swamp" in 1949 and 1953. Drill instructor Richard Hudson recalls a platoon entering a shallow marshland in 1953–54. The author witnessed only one invasion into a muddy marshland during two tours as a Parris Island drill instructor from 1953 to 1959. Moreover, swamp marches were not a part of the recruit training plan, and they required valuable time to clean a platoon's training and personal gear free of the thick tidal muck. Most marches were disciplinary in nature and left up to individual DIs. Few realized that a terrible tragedy might result from entering a Parris Island swamp or tidal stream.

DIs never deliberately ordered any event to cause a recruit's death, but poor judgments were made. However, until 1956, Parris Island's training record was one for which the Marine Corps and the Congress was proud. Moreover, no one has recorded a recruit killing a drill instructor. Motion pictures take great liberties.

Parris Island

Any sadistic personality is searched for in psychological interviews and in the drill instructor schools. Through the school and experience, Marines there hone their skills to such professionalism that they epitomize the Corps. Marine Corps historian Lynn Montross stated:

> *The drill instructor—or D.I., as he is known—has long been one of the proudest and most controversial Marine Institutions. He was not created. He has evolved, rather, from the drill sergeant of the days before World War I. But the D.I. is a lordlier figure;* [who] *is surrounded by an aura of divine right that the boot is not allowed to forget. It is his job to make a Marine out of a civilian in three months, and he is as dedicated as the chaplain.*

A DI's reputation was at stake with each platoon, and an instructor's fitness report could be marked accordingly. Similar to the demands placed on athletic coaches, the DI was expected to produce a winning team or platoon. Even after major reforms were ordered in 1956, platoons were awarded training "streamers" that were affixed to the top of a platoon's guide on staff for achieving outstanding training scores. Therefore, many drill instructors reasoned that to excel, the old and time-tested "head in the bucket" methods were necessary to produce outstanding platoons. And they did. The use of platoon streamers was later abolished, to perhaps reduce the intense platoon competition.

Other drill instructors and officers held the opinion that recruits should be schooled in the same manner in which they were trained—with an iron fist if necessary. There was also the opinion that all new Marines were to be molded so as to never diminish the proud and illustrious combat record of the Corps.

Recruits still came from all walks of life. Most were average and impressionable young men and women, who for various reasons wanted to be Marines. Their numbers included bed wetters, momma's boys, disillusioned John Wayne types and some hoodlums to whom judges gave the choice of joining the Marine Corps or going to jail.

Drill instructors had to contend with city hoodlums and gang members, thieves, con artists and deserters who now and then chanced the South Carolina swamps to desert from boot camp. Some boots mutilated themselves. One depot commanding officer Major General Thomas G. Ennis stated, "No matter what you can think of, what happens at a recruit camp, if you dig back in the records, you will find that some time or other the same thing has happened."

General Ennis recalled that from 1960 to 1962, several boots believed that if their rifle trigger fingers were missing they would immediately be

CHAPTER NINE

discharged from the Marines. The recruits placed their fingers against a rifle metal butt plate, jammed a dull bayonet down on their index fingers and sliced them off. The recruits were discovered as they threw their severed fingers underneath a barracks. One recruit thought that his finger would grow back! Others have offered slight variations of the true story, and the recruits were dishonorably discharged from the Corps.

Drill instructors as of April 1956 were as in the past seldom supervised by officers or senior NCOs. The drill instructor still enjoyed near unlimited authority and had keen insights in the psychology of producing Marines. Who would dare alter a system that, even with acknowledged shortcomings, had proven remarkably successful in two World Wars and a recent Korean War.

Recruits were subjected to a bizarre sense of drill instructor humor and endless hazing, intended to drive home the message of obedience and esprit de corps. For example, one 1954 DI ordered his problem recruits to sleep in trees because that was where "shitbirds" slept. All who traveled the main Parris Island Boulevard thought nothing of seeing such a spectacle, since this was Marine Corps boot camp. Other DIs did not permit recruits to write over the Marine Corps emblem on their personal writing paper.

It is erroneous to think that all recruits were maltreated. They were not, but could be. Hazing was common for all. Those who were maltreated considered any physical corrections as just punishment for their mistaken ways. Recruits were ordered to shave with issued safety razors whether they needed to shave or not. Failure to shave could result in two recruits shaving one another while supervised by a DI. Incorrigible city hoodlums could be ordered to wear their leather gloves while facing another recruit and to strike one another in the face. The lesson could be bloody, but corrected any inner city mentality problem quickly in its harsh way. Such direct and swift corrections also saved recruits the possibility of a court-martial or lesser notations in their service record book.

Packages of cookies and candies ("pogeybait") that were mailed from a girlfriend surely ruined a recruit's day. Nor was it wise to receive a lipstick-sealed letter from a boot's "teen age queen." Recruits were required to lick off the lipstick from an envelope in front of the platoon. A pregnancy and a hasty Parris Island wedding was another matter. The marriages were performed in an empty base chapel and were quickly done. Emergency recruit leaves were necessary. But some boots declined attending a close family member's funeral so as not to be set back in the training schedule. That possibility may have been the greatest of all recruit fears.

First Sergeant Charles Carmin passed on a story to the author that vividly demonstrates the one time authority and imagination of many DIs. Drill instructor Staff Sergeant Procacino's 1958 First Battalion platoon was preparing for a major inspection to include "junk on the bunk." Recruits had to meticulously display their equipment on their bunk (bed) in a prescribed manner. Mistakes were not tolerated. "Proc" held display schooling with his platoon, being that drill instructors were graded for a platoon's performance for such a significant event. Some drill instructors required their recruits to sleep on the barracks floor the night before a "junk on the bunk" inspection, so as to have all clothing and equipment displays near perfect for the inspecting officer to see.

Staff Sergeant Procacino previewed one misfit's gear that was a mess. The drill instructor in true fashion informed the recruit that he and his display looked like "a bunch of shit," and that the "shitbird" boot might as well be on display with his less than perfect arrangement. "Proc" ordered the boot to lie at attention on his upper bunk, seized the perplexed recruit's rubber name stamp and pad, and branded the lad on the center of his forehead in black ink. The boot was ordered to remain lying at attention in the middle of his display.

In those days, the upper bunks often sagged in the middle. Being the recruit was on the top bunk, the lad was not in easy view of a short inspecting officer. The young lieutenant had to stand on a locker box to see an upper bunk display. There, in all of his awkward innocence, was this recruit lying at attention in the middle of his disarrayed gear, with his name stamped in the middle of his forehead. The lieutenant had to exit the barracks to laugh at such an unimaginable scene.

Corporal John Brown informed recruits on rifle qualification day, that if they were excessively nervous, to report to him for a "nervous pill." Many did. The pill was nothing more than an All Purpose Capsule (aspirin), and it did the trick. The platoon was the top shooting platoon of the week. Corporal Brown later had a major role in Jack Webb's movie, *The DI*.

Despite careful drill instructor selections and an outstanding drill instructor school, mistakes and accidents happened. Only one deplorable and egregious event could have far-reaching consequences. That occurred when Staff Sergeant Matthew Charles McKeon (his last name rhymes with "hewn") led Platoon 71 on a night march into an unfamiliar tidal stream on Sunday, April 8, 1956. Six recruits drowned that night, and McKeon was court-martialed. The sergeant's initial sentence was reviewed and reduced, and the drill instructor was fortunate to receive the lesser sentence. McKeon was reduced in rank and was allowed to remain in the Marine Corps. But health and other reasons caused him to depart the service, to eventually retire as a civilian in Massachusetts.

Staff Sergeant McKeon was a good person who successfully served in the Navy during World War II. He saw intense combat in Korea, and he

CHAPTER NINE

knew that being a drill instructor would enhance his Marine Corps career. He should have known better than to commit a series of errors resulting in the death of six recruits. Moreover, McKeon lived with the tragedy for the remainder of his life. Strangely, most of the platoon members who attended an April 2006 Fiftieth Anniversary Parris Island Reunion respected their sergeant. Most would have invited McKeon to attend the event.

Base recruits at the time knew little about the Ribbon Creek drownings. But it was obvious to some that a terrible tragedy had occurred. Major Timothy J. Murphy (USMC Ret.) was then a Third Battalion boot in Platoon 105.

> *We started out with a Tech Sgt and two Sgt's as DIs on March 22, 1956. We ultimately graduated having had a total of 7 DIs. The Tech Sgt was relieved when he was seen giving extra added instruction to a Marine on the 3rd Bn obstacle course. The recruit joined at Parris Island very overweight, and was having extreme difficulty completing the course (this was not our first time on the obstacle course). The bayonet practice range was at the same site. The DI was observed by one of the Recruit Battalion Officers using a piece of rubber tire from the Bayonet course on the Marine's butt to inspire him. This is the only known reason for any of our Platoon DI's replacement.*
>
> *On the morning of April 9, 1956, Platoon 105 was marched to the 3rd Bn mess hall for breakfast. We recruits did not know what had happened. However, there was electricity in the air in the chow hall. Just that feeling when observing the DIs at their breakfast, and the hushed tones of conversation that something had happened. We were marched back to the platoon area, put at ease and the Senior DI told us that the Marines had died. He further advised us that there would be a lot of press people on the base and we could expect to be asked about the incident and our treatment. We were advised to tell them that we did not know what happened and that there was nothing wrong with our treatment. The next Saturday we marched to the Rifle Range to begin weapons training.*

It is not the purpose of this work to revisit the Ribbon Creek tragedy. That is accomplished in an excellent study by John C. Stevens, *Court-Martial At Parris Island: The Ribbon Creek Incident*. The opinions of recruits surveyed following the drowning disaster is enlightening when studying the pre–Ribbon Creek recruit training program.

About 27,000 former recruits were forwarded a "Recruit Training Questionnaire" by the Marine Corps about one month following the Ribbon Creek tragedy. The survey concerned the quality and fairness of their boot camp experiences at Parris Island and San Diego. The sampling of recruits surveyed included those who enlisted prior to World War II, as well as those who enlisted as recently as 1956. U.S. Marine Corps officers reviewed the replies that offered some interesting results.

It was concluded that "abusive language, hazing, and maltreatment" existed at both depots. Generally, those surveyed contended the practice of maltreatment was not intended to hurt anyone. Abusive treatment, both "verbal and physical" was reported to be more prevalent at Parris Island than at San Diego. It was added, however, that "the Parris Island men are more enthusiastic than those from San Diego about the quality of the training they received, about their recruit treatment, and about their senior DI when compared with other Marines."

The questionnaire indicated that Parris Island Marines were more satisfied with the quality of their training than those from San Diego, most notably in 1955 and 1956. Older Marines from both depots rated their training higher than the more recent recruits. The survey also revealed that a "preponderance of Marines think they were treated as a recruit should be," with an approval rate again favoring Parris Island, of 83 percent. Only 7 percent of the former recruits thought that their senior DIs were "poor," with most of the small body of criticism being directed toward the years of 1953–54. The latter finding is especially interesting since an official Marine Corps Drill Instructor's School was established in 1952.

Officer supervision at both depots was reported as "relatively infrequent," yet Parris Island was cited to have the most. Abusive language occurred slightly more at Parris Island, yet it was believed that cursing was not intended to debase and was mostly practiced by drill instructors to correct mistakes or to stress a point.

Similar results were reported concerning ridicule or discomfort directed toward recruits. Outright maltreatment was noted as "often" by 30 percent of Parris Island's recruits and 28 percent from San Diego, but punishment was considered as training for "retarded or recalcitrant recruits." The report further stated that the basic rights of boots were rarely infringed upon by their DIs. Perhaps the most revealing statistic was that combat Marines rated their recruit training in greater esteem than Marines who were not in combat.

The questionnaires revealed that most of the former boots were protective of their drill instructors and approved a continuation of the tough training. Some Marines stated that boot camp should be more intense.

Lieutenant Colonel Tom McKenney offers that he was

> *privy to a study that showed that DIs led all other MOS/duty groups in nervous breakdowns and divorces.*

CHAPTER NINE

I was aware, even as a recruit, of the pressure the DIs were under; and I was also aware that their officer superiors expected rules to be broken (E.g. "thumping"); it was a case of "just don't let me catch you." It was that conflict between pressure to produce and fear of getting caught that caused my DI to flip out and punish me and others for his mistakes.

Recruit recollections unquestionably record that hazing and maltreatment did not suddenly end following Ribbon Creek. Major Joe Featherston recalled his Platoon 351, First Recruit Training Battalion experience, from October 1 to December 31, 1956.

For some inexplicable reason I changed my mind about going into the USAF and focused on the Marine Corps. I headed off to Parris Island simply because it was the depot assigned to the "east of the Mississippi" recruits. I'd never heard of Parris Island and had only a basic understanding of the turmoil associated with the court-martial then underway resulting from the drowning of six recruits on a night disciplinary march into the swamps behind the rifle range butts. I regret that we were not issued the sunglasses reportedly provided to the San Diego recruits in our bucket issue, but it didn't seem to make that much difference.

I arrived aboard the Depot in the middle of the night after earlier debarking from a train in Yemassee, SC and being herded onto a bus for the final leg into the Parris Island base. There was no way of knowing what the surrounding area looked like and whether you were, in fact, on an island way out in the water, like Devil's Island. Tired, apprehensive, not yet terrified, disoriented and confused, we were herded around for various processing aptly described by other Marines. I remember some of the confusion but the only very clear elements were being marched to morning chow at about 0500 in civilian clothes, in the dark and stumbling over guy lines for a bunch of large tents staked out on the field somewhere. When we were halted at the mess hall we were treated to the sight of a half-dozen "prisoner-chasers" armed with riot guns marching several dozen prisoners to chow. It was pretty well pointed out to us that this was what we could expect if we broke any of the rules. I was then about 5' 9" and 129 pounds, very much a "feather-merchant."

The next week was a blur of processing, uniform and "bucket" issue, skills testing, threats, intimidation and some physical "hand's on" instruction by the three DI's. In fact, as I've recounted a number of times, I was then so scared that all bodily functions, like routine bowel functions ceased to occur. I can tell you that your body woke you up milliseconds before the DI hit the light switch in the squad-bay and you were out of your rack stripping the bedding before the electricity actually illuminated the bulbs. Fear will do that.

My next personal memory was returning from aptitude testing about the third day and being summoned from the (relative) safety of the squad-bay by one of the junior DI's (Sgt Strickland). After a period of me knocking (pounding) on the door jamb and reporting, and Sgt Strickland not hearing me, he got up in my face and demanded to know "what are you going to be when you get out of boot camp, turd?" I responded (after the aptitude testing) "Sir, an air traffic..." but I never got the last word out before I received a resounding slap across the face. There were several repetitions of that before I caught on and responded, "Sir, a Marine, sir!" More slaps, too slow! It turned out that the DI's had noticed that I could do facing movements and could march in a straight line so I was selected as the platoon guidon bearer. In those days, we drilled under the Landing Party Manual involving eight-man squads formed according to height with the tall guys up front and the short guys in the rear of the platoon when in a column of squads. That put me at 5'9" in front of the first squad at about 6'3". No problem when marching, big problem when double-timing.

One very dark morning about 0530 after 20 minutes of physical training, the DI ran us off into the darkness on the parade deck and he (I thought) remained behind. After several instances when the first squad ran up my legs I decided to ease over to the left a bit. That worked okay until the platoon began to creep past me. I then had the remarkable thought that I would simply turn left, cut across the parade deck and rejoin them on the other side. Needless to say, DI's see in the dark. I was relieved of my honorary post and busted back into the sixth squad not without certain colorful words.

I recall another event that involved the mess hall one evening. The protocol was that the DI's ate after the platoon had gone through the chow line and sat down to eat. When the DI was done, so were you. By the time he appeared on the parade ground the platoon had better be in formation reading their Guidebook for Marines and ready to go. That meant you ate with one eye on your tray and the other eye on your DI. Sometimes he would "play" with you by pretending to be done just to see you shovel your food down your throat as fast as you could.

This particular night, the duty DI came out of the mess-hall and berated us for eating like pigs and causing him to be embarrassed in front of other DI's. He told us if we wanted to eat like pigs he'd treat us like pigs. He commanded, "left face, get down on your hands and knees, forward, crawl." He then called cadence as we crawled back about two blocks to the barracks while we "oinked" very loudly. That was not enough. In a

CHAPTER NINE

"column of files from the right" we continued to crawl single-file into the barracks, up the stairs, down the passageway, down the other stairs, across that passageway and back up the stairs again, over and over, oinking all the time. After an hour of this, it was too much for one of the recruits who began to giggle, then snicker, then laugh uncontrollably. The DI landed on him immediately, pulled him out of the line and placed him on the upper bunk inside the squad-bay hatch where he was required to laugh at us as we crawled past. We, of course, glared at him with undisguised hate. When we were eventually secured and put into the racks for the night, and the lights turned off, the hapless recruit was still sitting there, in the dark, still laughing. When he tapered off, after hours and stopped laughing, the DI, from his room, would scream out, "I CAN'T HEAR YOU!" and the laughing would resume.

Another personal recollection involved the issuance of assorted sizes of field jackets in late-October as the weather got colder. (There was NO attempt to issue correct sizes).

I was in ranks as the DI called out the names of the platoon in alphabetical order. Each recruit answered "Here, Sir!" I was day-dreaming. When my name was called, I responded "Yo!" The DI, in apparent disbelief, yelled "Yo, Yo, what the @#$%^&* you think this is boy, the *&^%$#@ cavalry? Get up here!"

One of the more traumatic events for me in the training cycle was the bayonet training using pugil sticks, which were broom handles with large cylindrical pads on both ends. Recruits were required to attack each other using bayonet fighting moves such as thrusts, jabs, butt strokes, etc. You progressed from one-on-one to two-on-one, then three-on-two. One of the very basic precepts was to never back up. Attack with your body balanced always forward. Well, I backed up and went down on my butt while I was facing two opponents. My reaction was to pivot in a circle on my elbows and shoulders kicking out in all directions to keep the other guys at bay. It did not amuse the DIs. I was pulled aside, told to kneel down and a DI pounded me on the top of the helmet with a pugil stick actually knocking me out. When I came to, I was face down on the ground surrounded by three now very concerned DIs who were talking over their options. It seemed prudent to me to keep my eyes closed for another few minutes and see how things developed. It all ended well and I rejoined the platoon a few minutes later.

Sunday's "Church Call" was a truly religious experience because it brought a temporary lull in the attentions of the DIs who marched you to the chapel but had to turn you over to the clergy for an hour. Even agnostics found religion in boot camp.

Parris Island

Swimming (now called water survival) was a particular terror for me because I was essentially an inner-city kid and a non-swimmer. I was able to pass the swimming test only because one of the DIs stood behind me and screamed into my ear, "Gawdamn you boy, get in there and swim!" I did, by holding my breath all the way across the pool.

There were a number of other memorable events but I will comment on only one more. We joined a new junior DI while at the Rifle Range. We were billeted in squad tents with plywood floors all lined up in neat rows. Just as in the wooden WWII barracks, the traditional "firewatch" was kept by a series of recruits walking a two-hour post throughout the night. During my watch I heard a command "FIREWATCH!" I double-timed to the street and reported to the newly arrived (and generously lubricated) Corporal John R. Brown who immediately got my attention by burying his fist into my stomach and turning to walk away. This occurred while the aforementioned General Court-Martial of SSgt Matthew McKeon was still underway. Not a lot in "hands on" had changed at that point. Jack Webb subsequently noticed Cpl Brown during the shooting of the classic black and white film "The D.I." Brown served as a film advisor and had a speaking role in the movie.

Fifty years later I recount these events to illustrate that, despite the significant physical and mental challenges that recruits went through, and the departure of individuals that couldn't or wouldn't measure up, the end product was a continuation of the long line of Marines. This experience had a lifelong effect on me and I came out of the recruit depot a new very young Private to face the real Marine Corps. Leadership was mandated despite the young age and junior rank. A 19-year-old Corporal had a lot of authority, and tremendous responsibility, well beyond his civilian peers back on the hometown streets. The accumulation or international travel, technical training, combat training, leadership training and exposure to a myriad of experiences certainly helped shape and mature me well beyond normal expectations.

Transitioning from the original out-of-boot-camp MOS of 0800 (basic artilleryman), I quickly became a communicator (radio operator, wireman, radio chief and communications chief). Over a period of nearly 10 years I served in several artillery batteries and in Vietnam combat with an infantry battalion before being selected for the Warrant Officer Candidate School in 1966. Following The Basic School and being re-designated into an entirely new MOS in aviation support as a Warrant Officer, I spent the next 12 years in various squadrons, two Marine Aircraft Groups, and at a US Navy command where I was responsible for the logistics support of all helicopters in the Navy and Marine Corps, then as the field Inspector

CHAPTER NINE

General of the command as Director of the Aviation Support Improvement Group.

I offer the above to document that my particular military career took me from Private to Staff Sergeant, from Warrant Officer to Major and allowed me to retire at age 38 with 22 years of service. The training received at Parris Island, and the follow-on experiences over two decades resulted in selection for the Navy War College, a most prestigious consideration, beyond my expectations on the 'grinder' at Parris Island in 1956.

In addition, it set the framework for a 27-year civilian career that culminated with a six-year period as the Chief Operating Officer of Airbus Industrie's North American operations and Chief Executive Officer of their product support company leading to establishing my own aviation consulting firm that served the aviation industry for over ten years.

It's perhaps a trite phrase, but it just would not have happened without the recruit experience and the service-unique guidance of the Marine Corps drill instructors. I am privileged to remain in contact with my senior drill instructor after 50 years and I have tried to ensure that he knows just how important his efforts were to my successes.

Ribbon Creek hit the Marine Corps harder than almost any of its famed battles in World War II, claimed the *New York Times* on December 7, 1958. The tragic incident is a milestone in recruit training history, and the story has been studied at the recruit depot's DI Schools. Significant program changes following Ribbon Creek included more physical training (PT), officer supervision of drill instructors and the opening of the San Diego and Parris Island gates for the public to witness recruit training. A public graduation ceremony dwarfed those of the past.

Ribbon Creek saw the "Smokey Bear" campaign hat (cover) return to the Corps in 1956 to improve DI morale. Single drill instructors were removed from recruit living areas and were billeted in excellent and more private quarters for them. Other training changes included recruit "snitch boxes" to report any drill instructor abuse. Drill instructor Charles Carmin recalled "thump sheets":

> *Platoons were picked at random and given a sheet with questions that they were ordered to complete. Recruits were asked if their DI ever hit them, or if they had seen a DI strike a recruit. Did they feel they were harassed? Did the Drill Instructors curse at them? Was there "thumping" the boot camp synonym for maltreatment.*
>
> *I [Carmin] always held school with my platoons if I suspected that we were to get hit with thump sheets. In no uncertain terms the recruits were*

told how they should answer the questions. A platoon would get hit with the sheets at random or, if a recruit turned in to sick bay with suspicious injuries such as a split lip or severe sprains or bruises.

During my tour in the late 50s, I was hit three or four times with thump sheets and still managed to complete my tour on the field. I must have schooled my recruits well.

Drill instructors who did not obey the new guidelines were court-martialed and the careers of many excellent Marines were ruined. Marines obey orders, and that is a primary message of this book. But drill instructors following Ribbon Creek believed the old system worked, and that theirs was an historical obligation to the Corps to train recruits as they were schooled in the past

Many officers and drill instructors marked time thinking that the memory of Ribbon Creek would fade away. Hazing continued at the two depots, and the physical correction of recruits went underground until after the Vietnam War.

The rifle coach is a very essential Marine in recruit training.

CHAPTER NINE

"Snapping In," or dry firing the rifle in the process of rifle qualification.

A Parris Island recruit and his rifle.

Recruits work the range butts and pull targets for shooters.

A recruit's rifle must always be clean. Not to do so could mean having to sleep with not only his own rifle, but also those of several "borrowed" from fellow recruits.

CHAPTER NINE

In recent times female recruits undergo rifle training as the male recruits do.

A female recruit works the targets and butts for the shooter.

Chapter Ten

> BY TODAY'S STANDARDS, THE DI OF EARLIER YEARS
> DID EVERYTHING WRONG.
>
> MASTER SERGEANT CLAY BARROW

The *Parris Island Boot* reported in 1956 that the metal huts and wooden barracks that were inhabited by recruits of the 1940s and 1950s would be demolished, and replaced by modern permanent barracks. This was then the largest construction undertaking in the history of the depot that began in the male Third Recruit Training Battalion in 1958. Similar construction followed in the other battalions and at the rifle range in the ensuing decades.

The new barracks were no longer scrubbed down by recruits armed with brushes and buckets of sand and water as the huts and older wooden barracks were. Still, in the recruit depots, there was always an exception to any general rule. The brick structures were well lighted and large enough to practice indoor drill. The showers were spacious, and the newer buildings and classrooms were air conditioned. Former recruits were astonished by the presence of washing machines, dryers and television sets that were intended for instructional purposes. Duty drill instructor offices remained in each barracks. The brick mess halls resemble modern cafeterias.

Drill and the standard military subjects remained significant in boot camp. Physical Training (PT) was stressed even more. Exercises included sit-ups, pull-ups and push-ups, bends and thrusts, and a three-hundred-yard shuttle run. Recruits ran obstacle and confidence courses and competed in tugs of war and in other strenuous games. Eight-foot 450-pound steel logs were used for teamwork and to develop upper body strength. Other exercises included rope climbing and body-carrying, which simulated combat tests of strength. Recruits were instructed in a new form of bayonet fighting while training with pugil sticks. Hand-to-hand combat techniques were taught,

CHAPTER TEN

and some boxing smokers were held. Platoon competition in all of these events was spirited at battalion and regimental field meets.

Drill instructors retained the same demanding standards as in the past. The *New York Times* stated the DIs were "cold eyed," had "buzz saw" voices and viewed their recruits as a "blob of uselessness" that must be molded into the image of a Marine. The *Parris Island Boot* recorded the DI "is the person who destroys the recruit's civilian ego and individuality, makes him an automaton that reacts immediately to an order, and literally turns him inside out and upside down."

Drill instructors since the 1960s were selected by screening teams that visited the posts of the Corps. The average 1968 student was twenty-six-years-old, had eight years of military service, possessed a high school education, was married and had one child. Many were Vietnam War combat Marines.

The eight- and thirteen-man squad drills that were taught in the late 1950s were discontinued since the many movements required excessive instructional time. Perhaps the most noticeable change was the emphasis placed on DIs to lead by example rather than by an iron hand. A drill instructor school sign read, "No man is fit to command another who cannot command himself." Yet, in fairness to previous instructors, they were not without that philosophy during the so-called head-in-the-bucket days. Moreover, as noted, hazing and "thumping" in the 1960s continued underground.

The receiving barracks' yellow footprints appeared early in the 1960s. Drill instructors met their platoons several days after the receiving process was completed, rather than as soon as the boots arrived on the base. Although this policy freed drill instructors from many administrative and processing details, it also eliminated much of the shock effect that DIs capitalized on in the past.

Platoons now trained in series and typically numbered approximately seventy boots, yet others listed larger numbers during the Vietnam War. As in the 1950s, named platoons listed the "Pittsburgh Pirates," the "Pittsburgh Steelers" or the "Buffalo Bills." One Vermont group was designated the "Green Mountain" platoon. Macon, Georgia sent Parris Island a "Rebel" platoon. The training schedule was reduced from eleven or twelve weeks, to eight weeks in 1965. The reduced schedule was necessary to meet the demand for Marines in Vietnam.

The course emphasized marksmanship, drill and the obedience to orders that drill instructors taught in bizarre ways. Other class instruction included the prisoner of war code of conduct and the Uniform Code of Military Justice (UCMJ). Parris Island maneuvers continued at Elliott's Beach, but shifted more to Paige Field. Marine Corps history and tradition was taught, as was first aid, personal hygiene and sanitation, water survival, the bayonet and hand-to-hand combat skills.

Private Walter Anderson arrived at Parris Island via Yemassee in September 1961. He recalled twenty-six years later:

> *I cried that first night in boot camp. I was frightened and I felt alone. The first week in boot camp was the toughest week of my life, as it was for everyone. To someone who has not experienced Marine Corps recruit training, it sounds brutal, but it really wasn't. It was for a purpose. It drew us* [the recruits] *together. They* [the drill instructors] *weren't tough on just one individual. The DIs were tough on all of us.*

Edwin Apple enlisted in the Marine Corps in August 1961. On his bus trip from Los Angeles to San Diego, the driver informed Ed, "You don't have to do this." The ominous warning soon had meaning as the new recruits arrived at the San Diego Recruit Depot. The men were stripped of their civilian clothing and issued a yellow short-sleeved shirt that was emblazoned with the red letters "MARINE CORPS," to be worn with red shorts. Haircuts followed and Private Apple thought that all was going well.

The following morning Ed was assigned to a platoon with approximately eighty other boots that came from all walks of life. There were farm and city boys and "some were not so smart." The platoon was housed in large huts that were furnished with bunk beds. By now the drill instructors were forming the recent civilians to resemble some sort of an organization and yelling, "You are the sorriest SOBs that I ever had." Only twenty minutes or less was allowed for chow, where "the food was just thrown on our trays. The DIs rationed food for the fat boys." All were ordered to eat everything that was served, while seated at long mess hall tables. At times marching band music was played.

Large metal buckets were issued to the platoon, and Apple wondered, what was a bucket for? "I soon found out along with many other issues and events during the next twelve weeks." The boots were berated as "shitbirds" and "maggots" as the weeks went by. "But we were learning how to be Marines as insane as some of the training appeared." Marching took up much of the time along with physical conditioning and hand to hand combat training. The bayonet was taught with the use of the pugil stick. Private Apple favored running the obstacle course. Some letter-writing time was allowed and large wooden locker boxes were used as makeshift seats and desks.

CHAPTER TEN

Some recruits by the fourth training week were dropped from the platoon as others joined Platoon 367. Following their rifle training and qualification, Apple's platoon was assigned a week of various duties on the San Diego base. Ed worked in the main complex cleaning the barbershop and restrooms. When he later went for his final uniform alterations, his trousers did not fit. Since he had gained weight, Ed's drill instructors surmised that Apple had been eating sweets ("pogeybait") at the main complex. "My drill instructors were some kind of mad: 'Look what you are costing the Marine Corps.'" The fact was that Ed had acquired more body muscle that showed as weight.

Private Apple completed his recruit training and later married a Marine. Following his Marine service, Ed enlisted in the United States Air Force to retire as a Senior Master Sergeant, later taking on a successful civil service career.

Michael Carey came from a military family and his father was a U.S. Army Green Beret. Mike was an excellent marksman who had earned a black belt in the martial arts. The young man was also intelligent and in excellent condition. However, Carey was unprepared for the shock of Parris Island, and especially the screaming and seemingly insane remarks and orders that he heard and received from his DIs. The eighteen-year-old youth wondered how the F-word could be put into so many sentences.

Carey's journey from Boston to Charleston, South Carolina, was by plane. He and others arrived on the base by bus in the dark morning hours of July 12, 1961. Carey recalled the horror of the old World War II Hygienic Building, the haircuts and "delousing" and where he first saw his three drill instructors, who were assisted by other DIs. Carey recalled the yellow footprints that recruits post on. He knew that he could survive recruit training if he did what he was told, kept his mouth shut, "and repeated my favorite prayer."

"The sheer power and oppressiveness of the heat stayed with me," as did the odor of the salt marsh. Once in the old World War II barracks the drill instructors quickly made it clear who was the Alpha Dog, with maltreatment and stringent rules. Recruits were informed that the DI "was our God! And we were not to worship any false Gods but him."

The platoon quickly learned how to do thing the Marine Corps way— not to take more at the mess hall than one could eat and how to address mail. "You" was a forbidden word to address DIs, and "Sir" was required. Recruits shouted "ATTENTION" when a drill instructor entered a squad bay, and "GANGWAY" to clear any avenue for a DI. The raw recruits were roused out of their sacks by DIs tossing metal garbage cans about the squad bay, and banging the cans with swagger and nightsticks. Bucket issues and rifles were acquired, and Carey's platoon engaged in a strenuous physical training program.

The PT uniform was a bright red cotton shorts with a yellow stenciled USMC and emblem on one side. The shirt was "a bright yellow blended fabric tee-shirt with USMC in red letters across the front." There was also a red hat with sneakers resembling less than fashionable U.S. Keds. After he left boot camp, Carey recalled that he never saw a Marine wear the physical training uniform or use the tie-ties that recruits used to hang out clothing that was brush scrubbed on concrete wash racks.

The DIs were never satisfied. All kinds of drill instructor mind games were played. Whenever it was so hot that outside training was discontinued, Platoon 239 was schooled inside a steaming wooden barracks squad bay with no air conditioning. When Carey's platoon saw the base red flag go up to cease all outdoor strenuous activities, they dreaded what followed in the squad bay. However, because training time was precious, the drill instructors made the most of any opportunity to school the platoon.

Helmet liners were painted silver to reflect the heat. Carey thought the liners looked absurd, but wisely kept his opinions to himself. Should a recruit suffer a heat stroke or have to take an emergency leave, the greatest fear in boot camp was being set back to a junior platoon. Carey, as many before him stated that, "I would rather have killed myself than get 'set back.'"

Carey's platoon continued to shape up, and on one rare experience Carey participated in a large sweep of a salt marsh to locate an AWOL recruit. Deserting recruits were at times handcuffed to causeway palm trees as the Military Police used searchlights to apprehend other deserting recruits.

The platoon was billeted in new rifle range brick barracks, that "compared to our wooden barracks were pretty luxurious." Private Carey fired impressive rifle scores that were well above the 220 points required to be an expert. Swimming instruction was held at a large indoor rifle range pool and a week of assorted police and mess duties were in store the week after departing the range. Other training included inspections, a march to Elliott's Beach, more schooling, PT and drill.

Parents and guests visited the base graduation ceremony, and the recruits were permitted base liberty. Carey recalled, "I think I know a little how a man getting out of prison feels." The new Marine reported to Sea School after boot camp and served with a rifle company in Vietnam. Carey was badly wounded in combat, which ended his Marine Corps career.

Private Richard E. Marks reported to the depot in 1964. His training day began at 0430, and included the prescribed course of schooling and marksmanship. There was always drill. Platoons now sang marching songs. This practice was frowned upon by earlier drill instructors who

CHAPTER TEN

wanted nothing to do with anything that was associated with the Army and their marching songs. Marks recalled one chant went, "I don't want a teenage queen / All I want is an M-14." The boot recalled the sand fleas, periods of elation and depression and the early and "perpetual state of fear and shock."

Marks reported several maltreatment investigations of drill instructors, but defended his DIs. He was critical of his platoon series officers and referred to them as "schmucks." Several platoon fights were witnessed, and the Private quickly learned that "the best way to stay alive here is to keep my mouth shut." He also stated that all of the brutality you hear about in boot camp was "90% false," and that the other 10 percent is just a lot of hard work. Private Marks qualified as a marksman at the rifle range.

Marks's platoon was permitted to purchase Christmas presents at the Marine Corps Exchange, where volunteers wrapped the packages for recruits. The practice seems to have been one more change from policies of the past. Graduation was eagerly awaited as Private Marks completed a boot camp that was "a long series of tortures," yet so administered that he observed an "esprit de corps is building in each of us."

Private First Class Marks reported for advanced infantry training in January 1965, and volunteered for combat duty in Vietnam. In Vietnam he wrote, "I am here because I have always wanted to be a Marine and because I always wanted to see combat." The young man was not disappointed. Prior to February 1966, his combat action prompted Marks to compose the "Last Will & Testament of PFC Richard E. Marks." On February 14, 1966, approximately six weeks before he was scheduled to return to the United States, Marks was killed in action at the age of nineteen. He was buried in the Arlington National Cemetery on February 21, 1966.

Private Asa Baber was a 1960 recruit who never forgot Sergeant Danny Gross. The boot was cursed at and provoked by his DIs, and informed that if he "ever got out of boot camp it would be in a hearse or in skirts, because I certainly didn't have the makings of a Marine." Yet his drill instructors taught Baber many "things that later saved my life."

Baber turned out to be a good Marine, who met Sergeant Gross in Vietnam. The future writer greatly respected the DI, ranking Sergeant Gross with President John F. Kennedy. Both men were Baber's "role models" and influenced the direction of his life.

As the accounts of Private First Class Marks and Mr. Baber attest, boot camp remained a trying experience in the 1960s. The drill instructor remained the main cog in the machine. Nevertheless, former Marines who returned to the island in the 1960s detected changes in the program that ensued following Ribbon Creek.

Parris Island

Robert Alan Arthur witnessed there were many more officers and civilians on the depot than in World War II. The wartime boot noted the paved parade ground, the formal graduations and the DI campaign "Smokey Bear" hats. Others were surprised to see some DIs actually smiling and to hear marching chants. Words like Goon, Feather merchant, Socks, Clown, Meathead, Girls, Mobs, Maggot, Coolie, Dummy, Turd, Skinheads, Yardbirds, Rubber Socks, Mac, Pup, Knucklehead, Chicken, Bird Brain, Lad and Zombie now disappeared. Boots were, at least in public, only addressed as Private or Recruit.

Life magazine in 1965 sent former Private Art Buchwald to "Alcatraz East," to be reunited with his former and never-smiling drill instructor of 1942. The celebrated humorist recalled the "saber-toothed, man-eating Marine Corps drill instructors, whose main nourishment was chewing on the nerves of the raw recruits." The 1982 Pulitzer Prize winner stated, "No one who spent more than 24 hours at Parris Island expected to leave it alive." Yet, Mr. Buchwald detected significant changes in training policies in 1965. Gone was the term "yardbird," filling pockets with sand, recruits wearing buckets on their heads and the recitation of, "This is my rifle; I called it my gun / And that's why I'm standing bare in the sun." Buchwald also observed, as thousands had before and would after, that "The DI who is perceived of as a sadist in the beginning of boot training, usually winds up as a father figure before it is over."

In his classic style Art Buchwald humorously explained the enigma of Marine Corps boot camp that has to be endured truly to know the accomplishment of becoming a United States Marine.

Mr. Buchwald first met his drill instructor, Pete Bonardi, at 0400 one Parris Island morning, and was asked,

> "WHO IS YOUR FATHER?"
>
> "I told him *Joseph Buchwald of Queens, New York.*" "*NO HE ISN'T. I AM YOUR FATHER. AND I WILL BE YOUR FATHER FOR THE NEXT 10 WEEKS. IS THAT CLEAR?*" "*Yes, father,*" I said. "*NOT FATHER—SIR! YOU WILL CALL ME SIR AND ONLY SIR. DO YOU LOVE YOUR FATHER?*" "*Yessir.*" "*THEN YOU'RE GOING TO LOVE ME, BECAUSE I'M GOING TO LOVE YOU LIKE NO FATHER HAS EVER LOVED A SON. IS THAT CLEAR?*" "*Yessir.*"
>
> "*HOW MANY PUSHUPS DID YOUR FATHER MAKE YOU DO EACH DAY?*" "*None, sir.*" "*WELL, I'M GOING TO MAKE YOU DO 50 PUSHUPS A DAY. DO YOU KNOW WHY?*" "*No, sir.*" "*BECAUSE THAT IS WHAT A FATHER IS FOR.*

CHAPTER TEN

TO MAKE HIS SON INTO A MAN. YOU'RE A MISERABLE EXCUSE FOR A PHYSICAL SPECIMEN. ONLY A FATHER LIKE MYSELF COULD BELIEVE HE COULD MAKE A BAG OF BONES LIKE YOU INTO A MARINE. DO YOU LOVE ME?" "Yessir."

"GOOD, NOW HIT THE DECK AND START DOING THE PUSHUPS."

He stood over me. I collapsed after 10. "I THOUGHT YOU LOVED ME," he shouted. "I do sir." "HOW CAN YOU SAY YOU LOVE ME WHEN YOU WON'T GRANT ME A LITTLE FAVOR LIKE DOING 50 PUSHUPS WHEN I ASK YOU TO?" "I don't know, sir," I mumbled with my head in the dirt. "GET BACK IN LINE," he said in disgust.

It wasn't the first time I disappointed him. It seems every day I did something to upset our father-son relationship. I wanted to please him in the worst way, but I didn't know how. When I made my bed in the morning, he tore it up and threw the sheets and blankets on the floor. When I talked in ranks he made me scrub the head with my toothbrush. And when he didn't like the way I crawled on my stomach in the mud he made me march with a full pack and rifle around the barracks all nightlong. Every time he punished me for the slightest infraction he asked if I still loved him, and I always said I did.

On a 20-mile hike we were given one canteen of water to last us the entire trip. I drank mine after 10 miles. Bonardi was very sympathetic and told me if he wasn't my father he would give me more. But he loved me too much, and he would rather have me faint than spoil me. Then he kicked me in the pants. "NOW GET MOVING, OR ELSE I'LL KICK YOUR BUTT FOR THE NEXT 10 MILES."

Bonardi woke me in the morning and put me to bed at night. Then he would inspect my rifle and make me get out of bed and clean it again. And so it went. The 10 weeks seemed like 10 years because Bonardi never left my side. Then we said goodbye. I vowed I would forget him as soon as I left Parris Island.

That was years ago. Despite my vow, every Father's Day I seem to remember him. And what is really weird is that when I look back on the role he played in my life, I really do love him.

Women Marines play a significant role in today's Corps. Only Parris Island trains the female recruits.

A formation of the Women's Recruit Training Command (WRTC). The female battalion is presently identified as the Fourth Recruit Training Battalion.

CHAPTER TEN

Passing in review.

Female Marine recruits undergo the same exacting inspections as the males do.

The legendary Lou Diamond posing with Women Marines in 1945.

CHAPTER TEN

Parris Island

Modern quarters for female recruits and Marines are far superior to those of the past.

Female recruits exit the Parris Island gas chamber.

Chapter Eleven

BOOT CAMP...IS THE PRICE OF MEMBERSHIP IN A
PROUD FIGHTING FRATERNITY.

CORPORAL GILBERT P. BAILEY

Old habits die hard, and forbidden training practices continued at Parris Island following the Ribbon Creek reforms of 1956. Hazing and "thumping" remained at the two recruit depots as published books recorded the experience that was once Marine Corps boot camp.

William Mares's *The Marine Machine* (1971) recreated in words and photographs the making of a Leatherneck, as Mares followed a platoon through boot camp for ten weeks. The author was critical of persons who "only care if the recruits march in step and get to dental appointments on time." Mares quoted barracks training language, referred to "kicking ass and taking names" and stressed the point that boot camp was a training site for a more serious exercise that was played out in Vietnam. Some viewed Mares's book as another account of training brutality, but to former Marines, it confirmed that San Diego and Parris Island still produced loyal and disciplined Marines.

Richard Stack's *Warriors: A Parris Island Photo Journal* (1975) photographically reported recruit training from the moment the boots are informed, "As privates in the USMC you are the lowest of all ranks." The book also affirmed that despite reforms, the Corps continued to produce recruits from the same basic mold as in the past.

Former Marines-turned-novelists recorded the more physical part of the program. Robert Flanagan's *Maggot* is a 1971 novel with an abundance of profanity and a theme of "brutal" training that serves as a setting for two recruits "to make decisions that will shape the rest of their lives." Parris Island boot and Marine Corps Sergeant Edwin McDowell published a 1980 novel entitled *To Keep Our Honor Clean*. Several Vietnam War era motion pictures affirmed that the old training methods tenaciously remained, yet at times with an overactive Hollywood imagination.

The United States Marine Corps remains a no nonsense force. Marines train for war and the Corps prides itself on "Being The First To Fight." Marines are basically infantry. It is as simple as that.

Training to kill another human being is abhorrent to our society, but the world is not always a nice place. When creating "warriors," cursing, physical corrections and harsh endurance have been used notably in the U.S. Army and in the Marine Corps. The bottom line is victory and survival. Brutality toward another person should not be tolerated. But at times a fine line separates what some consider instructional maltreatment and training that is possibly misunderstood.

The Marine Corps in the 1970s indicated that everything was "well in hand." Senior officers announced plans for a "leaner, tougher Marine Corps," and in 1971 the *New York Times* and *Newsweek* magazine stated the Corps was not relaxing its standards concerning practices such as saluting, reveille and hair regulations, as the other military services did.

Yet, in March 1971, the *Times* stated the Corps was drifting away from the "traumatic ordeal" of recruit training for a policy of "low stress" that focused more attention on respecting "the dignity of the new recruit." The newspaper stated that the "old tradition of frightening raw recruits into obedience" was "wasteful," and that the policy should become a practice of the past. Conversely, alleged "permissive" training measures were not welcomed by many current and former Marines, who voiced their objections in publications such as the *Marine Corps Gazette*.

While the Marine Corps was eliminating the so-called and past unpopular recruit practices, Harry Jeffers and Dick Levitan published *See Parris And Die* in 1971. The book was hardly objective, as the title suggests, and the book received tepid reviews. The work focused on recruit maltreatment cases that set off a barrage of unfavorable criticism of the Corps. For example, the authors reported that in a twenty-seven-month period from 1964 to 1966, 120 drill instructors were relieved of duty at the two recruit depots and that 73 DIs were charged "for maltreatment or abuse." Military training deaths unfortunately occur. But the suggestion of seeing Parris Island to die was absurd.

The *New York Times* picked up the maltreatment theme in 1971. The newspaper reported in July that thirty-nine Parris Island boots were hospitalized for kidney problems after they engaged in strenuous exercises for thirty minutes. The *Times* stated in September that two Parris Island recruits died in training within twelve hours time, and later recorded the deaths of one depot sergeant and a private who were accidentally killed on the hand grenade range.

CHAPTER ELEVEN

In June 1972, *Atlantic* magazine attracted more attention to Parris Island when a conscientious objector reported his unfavorable experience as a recruit. That same year, the *New York Times* reported that recruit training was "better" and "tougher," while others demanded a "hard look" at the alterations in the recruit program and the reform of those methods that had successfully prepared combat Marines for years.

The training debate continued in 1974 and 1975. Former Marine William Hart challenged the Corps' claim that it builds men was a "myth." Others asked if enough good men could be found to become Marines in the socially turbulent 1970s.

As the American Vietnam experience ended, recruit training attracted additional press coverage in *Newsweek* magazine in 1976. It recorded that "an outbreak of boot-camp brutality, sloppy recruiting and a breakdown in discipline have shaken the esprit of the once strict-and-straight Corps." *Time* magazine embellished the argument by recording that the Marines were taking substandard men to maintain its authorized strength, causing greater difficulties than had ever occurred in the Corps before. For example, the journal claimed that in 1975 the AWOL rate of Marines "of 300 per 1,000 personnel was greater than that of the other services combined," and that the "desertion rate of 105 per 1,000 enlisted men was twice that of the other services combined." The article and others added the Corps was awarding more bad conduct discharges than the Army, and that Marine reenlistments were down.

The *Times* stated in 1976 that in five years previous, 13 Marine recruits died during the same period of time that the Army had 14 recruit deaths, even though the Army had more trainees. The paper added that twelve additional boots died from suicides, drowning, etc., and that in the four years previous, 249 Marines at San Diego and Parris Island were injured seriously enough to be hospitalized. The paper stated that from November 1, 1975, to June 15, 1976, 20 drill instructors at the two depots were convicted for maltreatment, and that the number of courts-martial for instructors had tripled from November 1975 to June 1976.

The two depots always regretted the death of a recruit, and all efforts are made to maintain an accident free program. However, training accidents occur. Heat strokes are always possible and accidents for many reasons take place. Murders and suicides have been known among recruits and permanent personnel, and some base visitor accidents occur.

The Corps' two recruit depots received more unfavorable attention during the mid-1970s than since the tragic Ribbon Creek incident of 1956. The notoriety reached firestorm proportions from two instances involving recruits in training at both Parris Island and San Diego, in 1975 and in 1976.

Parris Island

In what the *New York Times* called "one of the more bizarre episodes of abuse by drill instructors," Private Harry Hiscock was shot in the hand with an M16 rifle by a Parris Island DI in 1975. The charged sergeant maintained that he did not know the weapon was loaded and that he only wanted to scare the recruit. Although his claim was most likely true, the DI made an egregious error pointing a weapon in the direction of another Marine. Noncommissioned officers have been reduced in rank for an accidental discharge, even in war.

Far more damaging was the death of Private Lynn E. McClure. *Newsweek* magazine reported this tragedy was "the most troubling controversy since Sgt. Matthew McKeon led six recruits to their deaths in the tidal swamps of Parris Island, S.C., in 1956." Although journalistic accounts varied, it was reported the twenty-year-old Texas recruit attempted to enlist in the Army and the Air Force. He was accepted by the Marines and never should have been allowed to enlist. McClure's parents labeled him "mentally handicapped," and he scored only "seven out of a possible 100" points when he took the mental aptitude test. Press accounts referred to McClure as a "born loser and a high school dropout," who was arrested several times.

The San Diego recruit "got off on the wrong foot" from the very onset of his training, and was assigned to a problem recruit motivation platoon. The frail 115-pound McClure wanted no part of pugil stick training and refused to engage in combat with the padded poles. On December 6, 1975, it is alleged that his instructor would not accept such insubordination and ordered several larger boots with pugil sticks to attack McClure to force him to fight. The combative recruits took their orders in earnest and knocked McClure to the ground. He reportedly begged his assailants to stop striking him before lapsing into unconsciousness.

Private McClure never regained consciousness, and died in a Houston, Texas, Veteran's Hospital on March 13, 1976. The United States Congress called for investigations in Marine training practices. McClure's parents filed a $3.5 million lawsuit against the Marine Corps. The instructor who supervised the bayonet-type training was later acquitted on the grounds that the tragedy was a training accident. The episode, however, triggered "a trial larger than that of any individual," since the defendant in the case was once again, as in 1956, "the Marine Corps itself."

The *New York Times* stated that previous reforms in Marine recruit training had not sufficiently changed numerous abuses, and that "scheduled Congressional investigations at both Marine training centers should go forward without fail." By May 1976, Defense Secretary Donald H. Rumsfeld ordered a review of training methods in all of the armed forces,

CHAPTER ELEVEN

if for no other reason than any brutality could harm the reputation of the new all-volunteer American military experiment.

Other criticism of the Corps continued in the press. The *New York Times* reported in June 1976, that the Marines were losing as many as one out of every five boots for psychiatric reasons. *Esquire* magazine continued the brutality theme in September 1976.

The Marine Commandant, General Louis H. Wilson, appeared before the Military Personnel Subcommittee of the House Armed Services Committee on Recruiting and Recruit Training in the aftermath of the McClure case. More recruit training changes followed. Ensuing and new recruiting policies included stricter guidelines for Marine enlistments, a review of all recruiting methods and the abolishment of Marine recruiting posters that featured the legendary stern-faced DI. Recruit training hours were shortened, motivation platoons were to be abolished and an even larger number of officers were assigned to oversee recruit training.

The *Beaufort Gazette* of May 27, 1976, let it be known that Commandant Wilson would no longer tolerate any continuation of recruit abuse. The general stated, "The image of the fierce bellowing drill instructor standing nose-to-nose with a scared young recruit will disappear as part of an effort to end boot camp training abuse." Yet the Corps had its defenders as well. Congressman Robin L. Beard argued that significant training improvements had taken place in recent years. Former Marine officer and Congressman, Paul N. McCloskey Jr., insisted that basic training should be as "tough as combat itself," even if some persons were accidentally wounded or killed. Their arguments were blunted, however, as additional and unfortunate incidents occurred at the recruit depots as well as in the United States Army program.

By 1978 the Marine Corps placed even stricter regulations on its recruiters to enlist only desirable women and men. Yet, *Time* magazine reported that year that little had changed concerning recruitment, and that "some Marine recruiters will sign up almost anyone." By 1979 the seemingly fruitless efforts to draw quality men and women into the military services caused others to question if the draft should not be reinstated to induce the procurement of better people. However, Commandant General Robert H. Barrow was seemingly satisfied with the training reforms. In a speech presented to Parris Island's officers and noncommissioned officers at the depot theater on August 9, 1979, the general congratulated "the recruit training team for doing a good job—'the right way.'" The nation's press was soon satisfied, and once more supported the nation's need for a Marine Corps.

The state of affairs at Parris Island and San Diego continued attracting the attention of the press. However, charges were not splashed over the front pages of newspapers as they were since 1956, in part attributed to Marine Corps recruiting improvements and drill instructor supervision. Furthermore, the Marine Corps discharged undesirable recruits and refused to tolerate drill instructors who did not toe the line.

Because of continuing recruit training changes and improvements, the creation of an all-volunteer military force, and a diminishing memory of the protested Vietnam War, Marine Corps recruit training by the mid-1980s was once more in the approving grace of the public eye. Enlistees were required to meet higher educational standards, military pay and opportunities were more attractive, and greater opportunities were offered to female Marines. The women experienced much of the same training that was received by male recruits, in an ever-changing program that instituted a "Basic Warrior's Training Course" in February 1988.

Regardless of the post-1956 training changes, most boots considered their training demanding and tough. Private Frederick Shepherd recalled the yellow footprints in front of the receiving barracks and the cadre that addressed new recruits with a barrage of profanity in 1975. During the first several days the boots received their close-clipped haircuts, their hygienic and initial administrative processing and medical and dental attention.

Shepherd's platoon witnessed the traditional shock treatment. He noted that his drill instructors informed the boots what they should write home in their first mail. Any criticism of a drill instructor could now launch an investigation. Barracks instruction was conducted in a "school circle," and discipline exercises included recruits jumping in and out of their bunks and a punitive locker box drill known as the "squad bay 500," in which recruits pushed the heavy green wooden boxes up and down the squad bay until the boots tired. Shepherd recalled a small amount of drill instructor "thumping" done in out-of-the-way places, such as in the barracks' "heads."

Recalcitrant recruits were dispatched to the motivation platoon, where they were disciplined by having to perform such tasks as digging ditches and cleaning the grease traps of a mess hall cooking range. The planned disorientation of the platoon mostly disappeared as it visited the rifle range for three weeks. The range instruction was followed by one week of duty in a depot mess hall. The final training weeks were spent in additional instruction and preparation for a final inspection that marked the near completion of boot camp. By this time the recruits were awarded such liberties as additional smoking privileges and permission to purchase newspapers from vendors on the base.

Drill instructor humor remained. On Christmas Eve, Shepherd's DIs ordered the recruits to hang their socks from the ends of their bunks. When the platoon awoke on Christmas morning everyone's socks were empty.

CHAPTER ELEVEN

Shepherd did not view the prank as cruel and insensitive hazing, admiring his DIs and maintaining that boot camp was not all that tough.

Private Dana Fitzpatrick enlisted in 1981. Like Shepherd, he was air transported from the Atlanta, Georgia Armed Forces Induction Center to Charleston, South Carolina, and bused to Parris Island at night. The platoon met its DIs in their battalion squad bay after all processing was complete. The first weeks of training included schooling, drill and physical training and exercising in which the drill instructors took part. Rifle range training was reduced to two weeks.

Mess duty followed, and Fitzpatrick's platoon was divided to work in three mess halls on the base. The high point of Private Fitzpatrick's experience was the individual combat course. Other activities included field meets, pugil stick fighting and inspections. Competition was intense in the training series of four platoons. Fitzpatrick stated that his drill instructors were "all right" and that one was very religious and did not curse at all. He further observed that if one did what he was told to do, boot camp was not that hard. Base liberty was granted the platoon prior to graduation, after which Fitzpatrick received eighteen days of leave.

Female training remained much the same in the Vietnam War era. Private Barbara Cherbonneau recalled:

> *In August of 1973, I made a huge mistake and got married. It did not take long for me to realize my error, and begin to investigate a way to get myself educated and out of my predicament. I began visiting recruiters in Bennington, VT, and selected enlisting in the USMC. I came to this conclusion for various reasons, among them being that my Dad and my older brother were former Marines, and that the Marines had recently changed their rules, allowing married women to enlist. I was told by my recruiter that I was the very first married woman to enlist.*
>
> *So, in March of 1974, in the cold Vermont dark morning I went to Albany, NY to begin my journey to boot camp, where I met "Humphries" in the Albany Airport. Since we were not allowed to use first names in boot camp, to this day I do not recall her first name. During our flight, we compared recruiter stories and laughed over the discrepancies, wondering which tale would turn out to be closer to the truth. In the end, her recruiter told a more honest tale.*
>
> *We had been advised to bring pajamas, a robe, panties, bras and a few other "civilian items," the fewer, the better. Later I was frequently asked by guys if we were issued Marine green underwear. It felt weird not to bring any cosmetics or perfume and jewelry was limited to a watch and simple rings. At the Atlanta, Georgia, airport we found "more of us" and*

Parris Island

speculation on just what we had all signed up for was rampant. A long bus ride later we found ourselves in the dark and frightened at the Recruit Depot on Parris Island

Upon arrival at the barracks, I was delighted to see a Coca Cola machine. My happiness was short lived when we were told that recruits were not allowed to use it. The machine was there for the convenience of the DIs and if we ever tried to use it, they would KNOW, and heads would roll!

The clothes we were issued to wear as recruits were light blue huge legged shorts and huge blouses. I looked at myself in the mirror only once. The classes we attended included make-up—aha…that was why we were told not to bring any. We were issued a very good make-up kit during instruction but were allowed to only use the kit when wearing our dress uniform. Our English composition skills were evaluated when we were assigned to write autobiographies. Those whose skills were lacking in the area were assigned further instruction.

Our exposure to male DIs was limited to drill instruction. The men were firm, polite and to my knowledge never put their hands on any female recruit. I actually liked drill and looked forward to it…except for those pesky sand fleas. Later my marching skills came in handy when I received a college band scholarship, despite the fact that I could not play any instrument nor even read music!

Our female drill instructors were with us constantly, marching us to classes, the mess hall and back to the barracks. They also were there for field nights and everything else. Our DIs were very strict, disciplined and firm… and they were unsympathetic to tales of woe from recruits. To my knowledge, no one was physically harmed by any of the three DIs that we had.

Although I was convinced that all nine weeks of boot camp were akin to being in hell, 45 of the 60 something women who began in Platoon 3B that March of 1974, graduated, including myself.

Throughout my life, the training I received from the DIs at Parris Island has been invaluable. My favorite person at Parris Island was our Assistant Drill Instructor, Sergeant Peak. I liked and respected her most, because she managed to strike the perfect balance of teacher, mentor, superior and yes…even friend. To this day, once in a while I think of her and hope she is having a very nice life somewhere.

Current female recruits, though mostly kept separate from training male boots, must abide by the Marine Corps' policy of gender segregation during basic training, while following nearly the same schedule as the men. This includes firing M16 rifles and basic warrior training.

CHAPTER ELEVEN

Running has always been an important exercise in the Marine Corps.

Recruit rope climbing was known to recruits as early as World War I.

Should a recruit fall from this challenging obstacle, it would be a wet landing in the muddy pond.

Pull-ups, another of the many training exercises.

CHAPTER ELEVEN

Scaling the obstacle helps to build upper body strength as do pull-ups.

No recruit quits. The recruit training is intended to build mental attitude and physical strength and endurance.

A fear of heights is overcome in most instances at the Rappeling Tower.

CHAPTER ELEVEN

Self-defense combat training has on many occasions been part of the boot camp curriculum.

Lifting heavy and cumbersome padded metal poles teaches recruits teamwork and physical endurance.

Later recruits experienced the same metal log drill, but the uniform had changed.

Another of the challenging obstacle courses.

CHAPTER ELEVEN

Pugil Stick training brings recruits into close combative quarters.

The two Marine Corps Recruit Depots strive to achieve complete safety, but accidents have been known.

Enduring "The Crucible" at the end of recruit training.

"The Crucible." Marching toward the Parris Island main station and toward the designation "Marine."

Chapter Twelve

COMPARING THE RECRUIT TRAINING OF TODAY WITH THE TRAINING OF YESTERDAY IS A WASTE OF TIME. WHO WILL LISTEN? NOBODY! SO I SIT AND TALK TO MY BEER.

TOM BARTLETT

Hundreds of depot visitors prompted the *Boot* newspaper to state that Parris Island had become a tourist attraction. Entertaining large numbers of visitors furthered public relations and brought a better "public understanding of what Marine training is about." Civilians attending graduation ceremonies are permitted to tour recruit barracks and mess halls and are treated to receptions as part of a graduation exercise. The famous Parris Island parade field was reduced in size to accommodate visitor automobile parking. The former base officers' club was opened to visitors and newly graduated Marines. It must be remembered that the above practices can be changed or altered, and that inclement weather requires graduation ceremonies to be conducted indoors.

The drill instructor school increasingly stresses leadership and physical training in addition to teaching drill and basic military subjects. Future drill instructors were and are reminded that it was a "no-win situation if a drill instructor grabs a recruit." One San Diego drill instructor permitted his recruits to chew gum and to "turn the television on in the barracks to view a football game on the weekend." Perhaps his decision was a protest.

Former DIs reacted to gum chewing and television viewing with the anticipated alarm. Yet by the 1980s and 1990s, there was little doubt that earlier practices, as well as the near unquestionable authority of the DI, had gone the way of "the old Corps." The *New York Times* reported that the Corps was "softening the role of the hard-boiled drill instructor, who for so long had been 'a folklore figure in the Marine Corps.'" The newspaper added the DI was "quietly losing his primacy" in recruit training and that the Marines were seeking to "blunt the role of the drill instructor." Parris

Island's commanding general stated, "We just had too many incidents, too much completely unnecessary harassment that had nothing to do with being a good Marine."

After the Yemassee receiving station closed on June 30, 1965, recruits were flown to Charleston, South Carolina, and bused to the depot at night. Once on base, the boots were deposited at the receiving barracks, where a sign informed them this was the "Gateway to the Corps." Showers, haircuts and the like were received during a platoon's forming. The boots were then delivered to their battalion squad bay, where they sometimes met the battalion commanding officer, and the platoon DIs, who recited "The Drill Instructor Creed."

> DRILL INSTRUCTOR'S CREED
> THESE ARE MY RECRUITS. I WILL TRAIN THEM TO THE BEST OF MY ABILITY. I WILL DEVELOP THEM INTO SMARTLY DISCIPLINED, PHYSICALLY FIT, BASICALLY TRAINED MARINES, THOROUGHLY INDOCTRINATED IN LOVE OF CORPS AND COUNTRY. I WILL DEMAND OF THEM AND DEMONSTRATE BY MY OWN EXAMPLE, THE HIGHEST STANDARDS OF PERSONAL CONDUCT, MORALITY, AND PROFESSIONAL SKILL.

No form of maltreatment or even mild hazing was officially permitted during the 1980s and in the following decades. Punishment such as physical exercises could be administered according to defined regulations, but it is likely that sporadic and illegal corrections were done. Some drill instructors, rightly or wrongly, saw themselves as little more than "troop handlers." Yet boots no longer had access to "snitch" or "amnesty" boxes. Privates could speak to officers about an alleged abuse.

Recruit's clothing was serviced by the base laundry or at barracks washing and drying machines. New barracks heads (bathrooms) were furnished with toilet stalls rather than rows of exposed toilets and urinals. Former Marines were astonished to learn the reason given for installing air conditioning into recruit barracks in 1986, was "because the hot and humid conditions here [Parris Island] are detrimental to the physical and mental well-being of the recruits." Insect repellent and body deodorant were permitted, and excessive stress was officially a practice of the past.

Female recruits were advised not to bring excessive items to Parris Island and to pack lightly. "If it doesn't fit into your pocket or backpack, leave it at home." For those not heeding the advice, forbidden items were placed in storage until one completed boot camp. All drugs were confiscated, and

CHAPTER TWELVE

eyeglasses were prescribed at Parris Island. Permitted hairstyles for women were a bun or a French braid. During the 1990s, the women were required to qualify with the M16 rifle. Distaff drill instructors were authorized to wear the DI campaign hat.

Homesickness belonged to every decade. Some male and female recruits wanted to return quickly home from what they were about to endure. Boots in recent times are allowed to telephone their families upon their arrival on base. Many parents make it clear that it would be wiser to endure Parris Island than to come home to mom and dad. Most disillusioned recruits remained at Parris Island and San Diego to complete boot camp.

The Marine Corps emphasized core values at San Diego and Parris Island, to "enhance" all previous training qualities and to "transform" our youth into Marines. It was reported in November 1996 that the Marine Corps would implement a values program and that "all Marines will be issued a values card, depicting our core values." The card's purpose and instruction was to "perpetuate a Marine Corps of young men and women who are dedicated to God, Country, Corps, and Family."

Drill instructors leaned "more toward the role of the teacher and the father-figure than…the old type of drill instructor." It was reported, "In the Marines, it's training—not hazing" that counts. Some in the past did not separate the two.

Boot camp is capped by enduring "The Crucible," which was first tested by drill instructor school students in September 1996. The course deprives recruits of sleep and any barracks comforts such as showers and clean clothes. The physical demanding course strongly stresses teamwork. The Crucible completion becomes "the defining moment" in the making of a Marine as the recruit may now wear the Eagle, Globe & Anchor emblem, and is accepted as a Marine.

The completion of recruit training is also accompanied by a family day, base liberty, and an impressive graduation parade. One lieutenant colonel noted, "We're trying to teach them [recruits] how to be a Marine before they leave here." For persons wondering whatever happened to the more stressful and former recruit training, Marine Commandant Krulak stated, "We're not changing boot camp. We are changing the Corps."

Both recruit depots at San Diego and Parris Island are currently operating on a nearly mirrored schedule. Recruit spirits are high, and the depots offer that today's Marines are better than ever. That may be relative and debated when examining the rich history of the Marine Corps. But there is little argument that the two depots, and their drill instructors, continue producing Marines who have preserved the traditions of the Marine Corps.

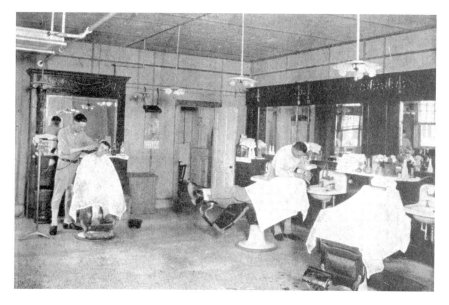

Few recruits ever saw barber facilities as lavish as these. The famous Marine Corps haircut was quick, and took away much of the individualism of a recent civilian.

Male recruits inside of the gas chamber.

CHAPTER TWELVE

Current water survival skills are much more a part of recruit training than in the past.

Parris Island

Recruit mess hall duty in boot camp. Such duties usually followed marksmanship training. Recruits could be assigned other duties as well.

Mascots have always been a part of the Marine Corps. The bulldog is usually displayed, although the breed is not usually considered as a war dog.

Chapter Thirteen

NO BOOT EVER FORGETS THE DAY HE RODE ACROSS THE PARRIS ISLAND CAUSEWAY AS A MARINE WITH A "SALTY" LEFT EMBLEM ON HIS "COVER."

LIEUTENANT COLONEL WALTER E. KIRACOFE

The evolution of the recruit training program has historically been debated and may forever be. For example, a 1918 Parris Island boot compared his experience as demanding as a cattle drive. Several World War II recruits offered that the postwar Parris Island boot camp was a joke: "The newest draftees going through there now don't see half the training we did, nor do they have the DIs hounding them like we did." Another added, "The famed boot camps for which we are famous are no longer what they were. They are comparative picnics." A 1943 recruit wrote that Parris Island had become a "boy's camp." He offered that the boots were "living in heaven with what we had there." Another opinion from a retired China Marine who later made Korea's Inchon landing and the Chosin Reservoir, and Vietnam offers:

> *Boot camp has to be a harsh experience. I feel the "old way" prepared Marines more so for the harsh realities of combat. I approve of The Crucible training concept, and if I had my way it would be a yearly event for all Marines, but should be run at least a week to ten days. Personally, I think more mental stress and harassment should be placed on recruits. If so, I believe that we would have fewer battle casualties if this was done.*

The most important function of drill instructors was and remains to instill discipline, endurance and esprit de corps. As noted, Marine Corps discipline has been in the past attained by tearing away civilian individualism, and rebuilding young men or women into a team member of the Corps. Philip N. Pierce and Frank O. Hough have skillfully written that since World War I, *the training course was deliberately made tough and exacting. The underlying motive was to effect a man's transition from civilian to military life as fast and as thoroughly as possible. Harsh disciplinary methods have sometimes been described as sadistic by horrified civilians who can claim no personal experience, nor have any real conception of what such training is all about. It is extremely significant that, with surprisingly few exceptions, men who have been through boot camp training are extremely proud of their achievement. Normal individuals emerge with a sense of accomplishment and a consciousness of having survived a trying experience together. Analysts attempt to establish the reasons which make Marine Esprit de Corps the extraordinary force that it is, invariably cite boot camp training methods more than any other factor.*

Others record that "Discipline is that which distinguished the Army from the mob—and the U.S. Marine Corps from any other military organization in the world." Yet the Marine Corps is not alone in recognizing the importance of discipline. Other military services as well as military colleges and academies are cognizant of the significance of discipline as well. However, the other military services approach to discipline has been different than with Marines.

Former Marine and author George McMillan once stated, "If there is a difference between Army and Marine Corps training it is in the *degree* of discipline imposed on its recruits by the Marine Corps." That discipline, according to McMillan, enabled World War II Marines to advance across hostile beaches while facing certain death. McMillan additionally recorded:

The difference between Parris Island and other military schools is in the quality of discipline (and discipline is not the same thing as brutality) it teaches and exacts. It is the Army man who boasts of the trials of physical endurance he has undergone in his training—the 40-mile hikes, for example. The Marine remembers the instant obedience and precision he was expected to show in a march around the block.

The Combat Correspondent had no intention of maligning the United States Army. Yet McMillan obviously expressed the opinion that the Marines as a smaller branch of service have placed more emphasis on disciplining their recruits than the much larger Army may or could have done in the past.

Numerous other Marines have recorded the significance of discipline as taught in Marine Corps boot camps. Lieutenant Colonel Wesley L.

CHAPTER THIRTEEN

Fox discovered, "I learned two things in boot camp, discipline and the fact that the noncommissioned officers (NCOs) ran the Corps—at least on the private's level." Former Marine and author Gilbert P. Bailey wrote, "Teacher, taskmaster and living example; guardian, guide and general manager of his platoon, these 'noncommissioned colonels' see to it that their men get 'the word' and 'the works.'" Bailey discovered that "to the DI, every act is a matter of supreme urgency," and to clean the head is regarded as "the most important military tactic since the battle of Saipan…[The DI] is a demanding perfectionist, a mountain of glowering impatience."

It remains incomprehensible to many civilians how the Corps' recruit discipline, hazing, and other past and present training practices instill the spirit to endure and provide a standard by which to measure future ordeals. Succinctly, Marines who completed the recruit training program believed there was little, mentally or physically, that they could not overcome. J.N. Wright referred to this characteristic in *Leatherneck* magazine in 1939:

> *The spirit of a "boot camp" is a peculiar thing. There we were, our civilian "independence" gone with the wind, blasted away by iron discipline, our bodies numb with weariness—and still, at the core of what we considered our misery, was a tiny spark of pride, a spark that assured us in a small, still voice that WE could take it, an assurance that no matter what new form of torture that drill instructor might devise, we could swallow it and still take more! We thought that little spark was our secret from the Corporal in charge, we guarded it jealously, little realizing that it was that same little spark that the instructor himself was purposely creating, that it was to build that spark that he was there.*

The stress, hostile environment, harassment and confusion of boot camp saw Marines not only through both world wars, but also through the rigors of the Korean and Vietnam Wars. Korea was a different kind of war that introduced so-called brainwashing and the exploitation of prisoners of war (POW). Major Gene Duncan, USMC (Retired), stated, "I think the fact that Marines fared so extremely well in the Chinese-Korean POW camps in the Korean War, and the fact that they did not succumb to the 'brain-washing' so effective on others substantiates," justifies the "harsh psychological treatment" of the past. Major Duncan also stated that "mental endurance—the ability to effectively function under great pressure—is every bit as important, if not more so, than physical endurance."

Parris Island

Robert J. Moskin agreed:

> *The Marine POW record had been clearly superior. When the United States Congress investigated the issue of prisoner of war conduct in Korea, the Senate report summarized: "The United States Marine Corps, the Turkish troops, and the Colombians as groups, did not succumb to the pressures exerted upon them by the Communists and did not cooperate or collaborate with the enemy. For this they deserve greatest admiration and credit."*

General William P. Battel recalled the importance of endurance in his recruit platoon, which was "proud of the fact" that their "training was rough, and that our drill sergeant was rougher than any of the others, and that we could take more than any of the other recruits in the other platoons." Lieutenant Colonel McKenney pondered, "I wonder why it is that, no matter how unpleasant boot camp was (and I hated almost every day because of the pressure on me from a DI I couldn't respect), in later years we all want to go back to Parris Island and remember and re-live it." It is interesting that most recruits of all times think boot camp should have been even tougher.

The importance of convincing troops that they are unbeatable was recognized by Commandant John A. Lejeune. The general stated that "the most vital thing is to make the men feel that they are invincible" and "that no power can defeat them." Years later, author Martin Russ wrote that World War II Marines were inspired to believe while in boot camp that they were "a select legion," and that a Marine must never let his comrades down. Russ, who respected Ernest Hemingway's evaluation of fighting men, added that the celebrated Hemingway once wrote, "I would rather have a good Marine, even a ruined one, than anything in the world when there are chips down." Robert Sherrod's 1944 book on the battle for Tarawa records the Marines there "fought almost solely on *Esprit de Corps.*" When Sherrod asked one Marine if he was afraid to assault Tarawa's beaches, the man replied, "Hell no, mister, I'm a Marine."

Author George P. Hunt wrote that the World War II Marines considered themselves unbeatable because they were indoctrinated in boot camp that they were "tougher than soldiers, sailors or the Japanese." Boots were also taught that a Marine must never "bring disgrace on the United States Marine Corps, whose proudest tradition was to accomplish any mission, regardless of the odds."

Drill instructors also taught their recruits there was no task the Army could do better than the Marines, although the Army had more men and material.

CHAPTER THIRTEEN

Recruits were indoctrinated with the belief that "A Marine sergeant is equal in training and experience to an Army lieutenant, and that a Marine private could take over an Army sergeant's job." The independent-minded Marines seldom wanted to imitate the Army by using "soldier" terms, and in the field it was considered sporting for Marines to rob Army supplies.

The Navy was a target for instilling a superior-than-thou belief among Marine recruits, too. Boots were told, and some believed, that one Marine was capable of taking on several "swabbies" at any time. The lesson dates back to at least 1933. That year *Newsweek* magazine carried a popular Marine verse:

> *Ten thousand gobs*
> *Lay down their swabs*
> *To lick one sick Marine.*

The unpublished poem's conclusion was, "Ten thousand more stood by and swore, it was the bloodiest fight they had ever seen."

The Marines' pride and unbeatable spirit that was learned in the boot camps was recorded by others as well. Colonel James A. Donovan wrote, "As in any military organization...as long as the men of the Corps consider themselves superior and elite," their performance will be superior. Another officer offered that if you teach a recruit that he is a superior fighting man, and "make him pay a price for membership in an organization with a reputation to uphold," he will uphold that reputation because it has become his reputation as well.

A First Marine Division chaplain observed such characteristics during the Korean War. "You cannot exaggerate about the Marines. They are convinced to the point of arrogance, that they are the most ferocious fighters on earth—and the amusing thing about it is that they are." Commandant Pate concurred. Drill instructors convince their recruits that, "Nobody in the world can whip them and that they can whip any two people their size or any size—and they go right on believing it and that's 50 percent of winning the battle right there."

Marine Corps' superiority is also drummed into recruits in their boot camp history classes. Lieutenant General Alfred H. Noble, recognized the importance of historical indoctrination and stated, "The inspiration of Marine Corps history on recruits is very marked. All somebody has to do is explain it to them, and they are right ready to receive it." Brigadier General Dion Williams commented that not only must men master their elementary military instruction in boot camp, "but also special instruction in the tradition and history of the Corps, in order that the young man joining the Corps may at the outset of his Marine career gain some of the pride and esprit in his chosen service."

Historian Bruce Catton offered the story of an American woman who visited a French hospital in World War I. She mentioned to one American patient, "Surely you are an American." The wounded man replied, "No ma'am, I'm a Marine." Most recently, an elderly World War II drill instructor and his wife traveled from Indiana to South Carolina by bus to attend a Parris Island reunion for former DIs. The elderly lady reported that her husband had recently recovered from a stroke, and as soon as he could speak, he stated that he was a Marine. Others during their final days have desired to revisit Parris Island and San Diego. Ashes of deceased Marines have been spread over Parris Island and most likely San Diego as well.

It has been this author's good fortune and privilege to meet and know many notable Marines during my enlistments and my writing projects. I have chatted with commandants and other general officers, and sat as a speaker with the youngest Marine Medal of Honor recipient. I have been honored to know combat seasoned and other Marines who in civilian life have made the Marine Corps proud. Considering all of their accomplishments and their many laurels, Marine conversations in most settings always return to boot camp and the drill instructors who made life miserable for the former recruits.

Lieutenant Colonel Tom McKenney USMC (Retired) served with and became a friend to Major General James L. Day. The general frequently enjoyed relating his San Diego recruit experiences to others. Colonel McKenney offers the following story to illustrate the indelible memories that Marines have of their recruit training, no matter their future station or rank.

> *James L. Day enlisted in the Marine Corps in 1943 and went through boot camp in San Diego; he is probably unique in that in later years he served as the commanding general of the San Diego base. General Day also earned the Medal of Honor as a corporal on Okinawa in 1945, but received the medal 53 years later as a retired major general. Jim Day had a great sense of humor, had many stories to tell and he loved to share them. One that was often shared was a boot camp experience. All laughed over it each time the tale was told.*
>
> *The Story: During boot camp Private Day's senior DI called Jim into the DI room and ordered Day to go to the base laundry and pick up the drill instructor's cap covers. These were the days of starched khakis, and the frame cap covers were removed when dirty, washed, starched, pressed and put back on the cap. For obvious reasons each Marine had two cap covers; and the DIs usually had several.*

CHAPTER THIRTEEN

As Jim stood at attention before his DI he was told, "Now Day, I want you to walk straight to the laundry, pick up my cap covers, and immediately return. You are not to go anywhere else. Do you understand me?" Jim replied, "Sir, oh yes sir! Private Day understands, sir!" The drill instructor added, "And Day, you are not to go to the PX!"

Private Day enjoyed his newfound freedom as he passed the PX on his journey to the laundry. Yet, Jim felt the temptation to at least have a sampling of all of that forbidden candy and ice cream in the PX. But, and as ordered, the recruit went straight to the laundry and was given the drill instructor's cap covers.

On his return trip Jim was again tempted, thinking of all of the candy and other forbidden fruit in the PX. The temptation was now even greater. He stopped, again thinking about all of the off-limits sweets awaiting him in the off limits PX. Private Day was overcome with his yearning for ice cream and disobeyed his drill instructor's orders.

Still more nervous—one might say frantic—Day started back to his barracks. Feeling a huge burden of guilt, and terrified at the thought of being caught in his misdeed, he hurried on. Day thought only of making it back to his barracks without being noticed or detained, so as to safely resume his normal recruit activity. Back in his hut he heaved a great sigh of relief, and resumed working on his gear. He had made it! Or, so it seemed for a few brief minutes, until the drill instructor summoned him.

DI; Day, did you go anywhere besides the laundry?
Jim (into his lie way too far to turn back): Sir, no sir; Private Day would never disobey the drill instructor, sir. Private Day went straight to the laundry and immediately returned, sir.
DI: Day, did you go into the PX?
Day: No sir! Private Day would not go into the PX, sir.
DI: Then how do you explain the fact that my cap covers were found in the PX?

Those of us who have been recruits can imagine the rest of the story. Private Day took his punishment, survived boot camp, and went on to serve nearly 50 years as a Marine, including close infantry combat in three wars. He received many decorations, including the Medal of Honor, three Silver Stars, two Bronze Stars, and seven Purple Hearts. But General Day never forgot his one boot camp violation, and he always believed that his indiscretion had kept him from being the platoon guide and not being promoted to Pfc out of boot camp.

Parris Island

Like all of us, as the general grew older he told the cap cover story many times. The long ago events in boot camp became increasingly important to him. As a major general he thought about that boot camp day and wished that he could re-visit those days with his old DI. Because of his high rank and position, the general easily located his drill instructor, and was relieved to learn that the elderly DI was still alive. General Day got the drill instructor on the telephone, hoping to relive with him that long-ago boot camp incident; but to Jim's enormous disappointment, the DI remembered neither Jim nor the matter of Private Day and the cap covers at the PX.

It is easy to forget that each of us had only one Senior DI, but the DI had many, many recruits.

It is astonishing, how so many former recruits compliment, praise and in some instances even love their former DIs. The testimonials are numerous. Commandant Robert H. Barrow stated, "My drill instructors lighted a fire in me 40 years ago. It's still burning and will until I die." Commandant Randolph McC. Pate offered, "The fighting spirit of the Corps could be traced to the drill instructors and their teaching." General Gerald C. Thomas candidly stated, "I'd never trade my boot training for anything in the world." Others concur.

The *Parris Island Boot* expressed the sentiments of former recruits in 1967. The base newspaper asserted that "after graduation and through life many recruits will appreciate the fact that they owe a debt of gratitude to their hard working and conscientious drill instructors," that infused a form of discipline in them that has unequivocally been essential to the Marine Corps' success.

Marine Staff Sergeant Norval E. Packwood praised Sergeant A.B. Suggs, "My own DI, who has been the inspiration for every cartoon on Boot Camp I have ever drawn." Author Herb Moore wrote of drill instructor Walter Egge, "For his leadership and courage, I am grateful. For the less obvious yet more encompassing set of values he instilled in me, I shall forever be in his debt."

Author Robert E. Shirley adds, "The tenacity I learned in Parris Island enabled me to earn three college degrees. I've had four careers in electrical engineering, accounting, corporate finance, and securities brokerage…My Parris Island training helped me assume positions of leadership in my church, cub scouts, condo association, and the Marine Corps League." Author Eugene B. Sledge praised his drill instructor Corporal Doherty: "To him more than to my disciplined home life, a year of college ROTC before boot camp and months of infantry training afterward I attribute my ability to have withstood the stress of Peleliu."

CONTRIBUTORS LIVING AND DECEASED AND CREDITS

(Rank and titles are not listed)
Alexander, Phyllis
Alvarez, Eugene
Anderson, Walter
Apple, Edwin
Arthur, Robert A.
Baber, Asa
Bailey, Gilbert P.
Barrow, Clay
Barrow, Robert H.
Bartlett, Tom
Battell, William P.
Berkeley, James P.
Bethel, Ion M.
Betz, Don
Billings, Vincent
Bonardi, Pete
Bowden, Ralph T.
Brannen, Carl A.
Brown, John R.
Buchwald, Art
Campbell, Elvis
Canedo, Billy
Carey, Michael
Carmin, Charles
Catton, Bruce
Cherbonneau, Barbara
Churchill, Walter A.
Coley, Perry N.
Corbin, Roy E.
Cosman, Samuel

Cox, O. T.
Creecy, Wilburn C.
Cruse, Bernard William Jr.
Darnell, Jim T. "Bulldog"
Dawes, Henry T.
Diamond, Leland "Lou"
Donovan, James A.
Duncan, Gene
Edwards, George T.
Ennis, Thomas G.
Farrell, Walter G.
Featherston, Joe
Fields, Lewis J.
Fisher, Carter
Fitzpatrick, Dana
Fitzpatrick, Mary-Ann (Perry)
Fox, Wesley L.
Gallant, T. Grady
Gibson, Hoot
Ginther, Julius J.
Greene, Wallace M. Jr.
Guest, James C.
Gulley, Bill
Hardy, Carle E.
Hart, William
Heinl, Robert D. Jr.
Hemingway, William E.
Hetzner, Don
Hornstein, Joseph
Hough, Frank O.
Hughes, Charles B.

CONTRIBUTORS LIVING AND DECEASED AND CREDITS

Hudson, Richard
Hunt, George P.
Johnson, Joseph E.
Jones, Katherine M.
Jordahl, Russell
Kiracofe, Walter E.
Kozak, Arthur
Krulewitch, Melvin L.
Kvaternik, Raymond G.
Lange, Louis
Lear, Warren F.
Lejeune, John A.
Leckie, Robert
Luckey, Robert B.
Manchester, William Jr.
Manchester, William Sr.
Mares, William
Mark, Donald
Marks, Richard E.
Mason, Don
McKenney, Thomas
McKeon, Matthew C.
McMillan, George
Montross, Lynn
Moore, Herb
Moore, Billy
Moskin, Robert J.
Murphy, Timothy J.
Noble, Alfred H.
Ogle, John C.
Oxford, John
Packwood, Norval E.
Paige, Mitchell
Pate, Randolph McC.
Patten, Thomas H., Jr.
Peatross, Oscar F.
Pfeiffer, Omar T.
Pierce, Philip N.
Pollock, Edwin, A.
Procacino (SSgt)
Puller, Lewis B. "Chesty"
Redenbaugh, Joan

Robinson, Ray A.
Rogers, William W.
Rupertus, William H.
Russ, Martin
Ruttle, L.F.
Schilt, Christian F.
Seymour, Harold K.
Shaw, Henry I.
Sheffield, Theodore M.
Shepard, Lemuel C. Jr.
Shepherd, Frederick
Sherill, James
Sherrod, Robert
Shirley, Bob
Silverthorn, Merwin H.
Sledge, Eugene B.
Smith, Robert A.
Smith, Larry
Snedeker, Edward W.
Stevens, John C. II
Stevens, John C. III
Strickland (Sgt)
Strong, Robert L.
Stuckey, Wilbur H.
Talpalus, Theodore S.
Thomas, Gerald C.
Throop, Homer A.
Trombly, Oliver
Truesdell, Donald L.
Tyler, Joe B.
Underhill, James L.
Vassey, Charles A.
Warner, R. R.
Watson, George
Webb, Jack
Williams, Dion
Wilson, Louis H.
Wilson, Walter
Winfield, Samuel
Wise, Stephen R.
Wright, J.N.

SUGGESTED BIBLIOGRAPHY

Alvarez, Eugene, PhD. *Parris Island: The Cradle Of The Corp.* Privately published. No longer in print.
———. *Parris Island: Images Of America.* Charleston, SC: Arcadia Publishing, 2006.
Bailey, Gilbert P. *Boot: A Marine In The Making.* New York: The Macmillan Company, 1944.
Champie, Elmore A. *A Brief History Of The Marine Corps Recruit Depot Parris Island, South Carolina, 1891–1962.* Washington, D.C.: Headquarters, U.S. Marine Corps, 1962.
Emerson, G.A. *With The United States Marines At Marine Barracks: Parris Island, South Carolina.* Brooklyn, NY: Albertype Co., n.d.
Flanagan, Robert H. *Maggot.* New York: Paperback Library, Inc., 1971.
The Hat: A Salute To The United States Marine Corps Drill Instructors, Parris Island, South Carolina. Swansboro, NC: Talent Associates International, 1997.
Helms, E. Michael. *The Proud Bastards.* Port St. Joe, FL: Karmichael Press, 1996.
Hunt, George P. *The Story Of The U.S. Marines.* New York: Random House, 1951.
Jeffers, H. Paul, and Dick Levitan. *See Parris and Die.* New York: Hawthorn Books, Inc., 1971.
Leckie, Robert. *Helmet For My Pillow.* New York: Bantam Books, 1979.
Mares, William. *Marine Machine.* New York: Doubleday & Co., Inc., 1971.
McDowell, Edwin. *To Keep Our Honor Clean.* New York: The Vanguard Press, 1980.
Moore, Herb. *Rows Of Corn: A True Account Of A Parris Island Recruit.* Orangeburg, SC: Sandlapper Publishing Co., 1983.
Packwood, Sargent Norval E. *Leatherhead: The Story Of Marine Corps Boot Camp.* Quantico, VA: Marine Corps Association, 1951.
Russ, Martin. *Tarawa: Line Of Departure*: New York: Zebra Books. Kensington Publishing Corp., 1975.
Shirley, Robert B. *Parris Island Daze. My Drill Instructor Was Tougher Than Yours.* West Conshohocken, Pennsylvania: Infinity Publishing, 2006.
Smith, Larry. *The Few And The Proud: Marine Corps Drill Instructors In Their Own Words.* New York: W.W. Norton & Company, 2006.

SUGGESTED BIBLIOGRAPHY

Stack, Robert. *Warriors: A Parris Island Photo Journal*. New York: Harper & Row, Publishers, 1975.

Stevens, John P., III. *Court-Martial At Parris Island: The Ribbon Creek Incident*. Annapolis: Naval Institute Press, 1999.

ARTICLES

Krulak, Lieutenant General Victor H. "This Precious Few: The Evolution Of Marine Recruit Training." *Marine Corps Gazette* April 1982: 48–55.

Warner, R.R. "Bucket, Steel, Galvanized," *Leatherneck*, March 1984, p. 51. *Leatherneck*, March 1946, p. 12 pictures Parris Island recruits with buckets on their heads.

ABOUT THE AUTHOR

Professor Eugene "Gene" Alvarez has written several Parris Island histories in addition to other books. Gene served in the Marine Corps from 1950 to 1959 and was honorably discharged with the rank of Staff Sergeant. He served with the First Marine Division in the Korean War, and was twice a 1950s Parris Island drill instructor. Alvarez received his PhD in history from the University of Georgia, and retired Professor Emeritus (Macon State College) from the University System of Georgia.

Semper Fidelis!

Please visit us at
www.historypress.net